Proceedings of the 2010
Complex Systems Modellin_

# CoSMoS 2010

Susan Stepney, Peter H. Welch,
Paul S. Andrews, Adam T. Sampson,
Editors

# CoSMoS 2010

Luniver Press
2010

Published by Luniver Press
Frome BA11 6TT United Kingdom

British Library Cataloguing-in-Publication Data
A catalogue record for this book is available from the British Library

CoSMoS 2010

ISBN-10: 1-905986-27-0
ISBN-13: 978-1-905986-27-9

While every attempt is made to ensure that the information in this
publication is correct, no liability can be accepted by the authors or
publishers for loss, damage or injury caused by any errors in, or omission
from, the information given.

# Preface

Building on the success of previous CoSMoS workshops, we are pleased to be running the third CoSMoS workshop co-located with the 12th International Conference on the Synthesis and Simulation of Living Systems (Artificial Life XII), in Odense, Denmark. The Artificial Life conference is an especially good fit for the CoSMoS workshop, examining critical properties of living and life-like systems and attracting a broad range of interdisciplinary researchers. The systems examined by these researchers are inherently complex, and various modelling and simulation techniques have become key to exploring and understanding their properties.

The genesis of the CoSMoS workshop is the similarly-named CoSMoS research project[1], a four year EPSRC funded research project at the Universities of York and Kent. The project aims are stated as:

> The project will build capacity in generic modelling tools and simulation techniques for complex systems, to support the modelling, analysis and prediction of complex systems, and to help design and validate complex systems. Drawing on our state-of-the-art expertise in many aspects of computer systems engineering, we will develop CoSMoS, a modelling and simulation process and infrastructure specifically designed to allow complex systems to be explored, analysed, and designed within a uniform framework.

As part of the project, we are running annual workshops, to disseminate best practice in Complex Systems modelling and simulation. To allow authors the space to describe their systems in depth we put no stringent page limit on the submissions.

We are delighted this year to welcome philosopher of science **Paul Humphreys**, Professor of Philosophy, University of Virginia, USA as our keynote speaker. Humphreys is author of the book *Extending Ourselves: Computational Science, Empiricism, and Scientific Method* that introduces the concept of *computational templates*, which form the core of many computational models. Our proceedings start with an extended abstract for Humphreys' keynote that expands on computational templates and asks how the same template can be successfully used on different subject matters.

The main session of the workshop is based on four full paper submissions:

---

[1] The CoSMoS project, EPSRC grants EP/E053505/1 and EP/E049419/1, http://www.cosmos-reseach.org

**Garnett, Stepney, Day and Leyser** describe an example of how to incrementally change a biological simulation (auxin transport canalisation in plants) in a principled manner using the process derived as part of the CoSMoS project.

**Ghorbani, Ligtvoet, Nikolic and Dijkema** investigate institutional frameworks to analyse socio-technical systems and understand complexity in agent-based models, providing a Kauffman model example.

**Polack** focusses on arguing validation of simulation in science, proposing the use of validity arguments across many validation approaches to provide evidence that simulations are scientifically fit for purpose.

**Stevens** presents an adaptation to Gosper's hashlife algorithm for the application to a three-dimensional kinematic environment, by simulating the environment which is modelled using cellular automaton rules.

For the first time at a CoSMoS workshop, we also invited authors to submit abstracts, for presentation in a poster session. Abstracts for the following posters are presented in the proceedings:

**Andrews, Ghetiu, Hoverd, Owen, Sampson, Warren and Zamorano** provide a group reflection on researching complex systems, providing an overview of general issues that can be both interesting but challenging.

**Araujo, Bentley and Baum** show how a simulation of chromosome missegregation in cancer therapies can provide new insights into cancer progression.

**Jones, d'Inverno and Blackwell** present an overview of their work modelling the haematopoetic cellular system, including how adopting an agent-based technique can facilitate a system-level conceptualisation of the domain.

Our thanks go to Paul Humphreys for presenting his keynote and to all the contributors for their hard work in getting these papers, abstracts and posters prepared and revised. All submissions received four reviews, and we thank the programme committee for their prompt, extensive and in-depth reviews. We would also like to extend a special thanks to the organising committee of Artificial Life XII for enabling our workshop to be co-located with this conference. We hope that readers will enjoy this set of papers, and come away with insight on the state of the art, and some understanding of current progress in Complex Systems Modelling and Simulation.

# Programme Committee

Paul Andrews, University of York, UK
Fred Barnes, University of Kent, UK
James Bown, University of Abertay, Dundee, UK
James Dyke, Max Planck Institute for Biogeochemistry, Germany
George Eleftherakis, CITY College, International Faculty of the University of Sheffield, Greece
Philip Garnett, University of York, UK
Nic Geard, University of Southampton, UK
Simon Hickinbotham, University of York, UK
Andy Hone, University of Kent, UK
Tim Hoverd, University of York, UK
Sara Kalvala, University of Warwick, UK
Adam Nellis, University of York, UK
Nick Owens, University of York, UK
Fiona Polack, University of York, UK
Simon Poulding, University of York, UK
Mark Read, University of York, UK
Susan Stepney, University of York, UK
Jon Timmis, University of York, UK
Alan Winfield, University of the West of England, Bristol, UK

# Table of Contents

## CoSMoS 2010

x

# Some Relations between Formal Structure and Conceptual Content in Simulations

Paul Humphreys

Corcoran Department of Philosophy, University of Virginia, USA
pwh2a@virginia.edu

## 1  Wigner's Question

In Eugene Wigner's article 'The Unreasonable Effectiveness of Mathematics in the Natural Sciences' [9] he claimed that "... the enormous usefulness of mathematics in the natural sciences is something bordering on the mysterious and ... there is no rational explanation for it." I want to show here that much of what Wigner found to be "unreasonable" has a straightforward explanation. We shall see that there is a number of reasons why parts of mathematics are applicable to the natural world and that these explanations differ in interesting ways. Unlike Wigner, I shall not restrict my attention to the natural sciences because the problem of applications stretches across the disciplines. I shall also switch the focus from traditional mathematics to computer simulations because a similar question arises for those techniques.

In [4, 5], prompted by some remarks of Richard Feynman, I introduced the idea of a *computational template*. In areas outside complexity science, these are often derived from *theoretical templates*. A theoretical template is a general representational device occurring within a theory, containing schematic, second order, property variables and such that, when all of the schematic variables have been substituted for, can be successfully used to represent a variety of different phenomena within the domain of that theory. Well-known examples of theoretical templates are Newton's Second Law, which in its simplest one dimensional form is $F = md^2x/dt^2$, where $F$ is the schematic variable; the mathematical theory of probability, within which the schematic variable $P$ requires the substitution of some specific probability measure; Schrodinger's equation $H\Psi = E\Psi$, where the Hamiltonian operator $H$ and the state function $\Psi$ are the schematic variables, and Lagrange's equation $I = \int_t^{t'} L\, dt$. These are theoretical templates because they are interpreted formulas occurring within the framework of a recognizable theory that is about some specific, albeit often abstract and very general, subject matter.

Not all substitution instances of theoretical templates are computationally tractable, whether analytically or numerically, but if the resulting, more specific, syntactic equation form has that property then we have a *computational template*. A computational template is a purely formal object; even if it originates as a substitution instance of a theoretical template, the original interpretation is removed to leave a syntactic object that has only a mathematical interpretation. Computational templates form the core of many computational models and one of the remarkable things about them is that they are often applicable across a wide range of subject matter, not only within a given science but across different sciences. So, once we recognize the existence of computational templates, we immediately face a different question than Wigner's, which is: How is it that exactly the same template can be successfully used on completely different subject matters? Taking an example that is close to the domain of complexity theory, the Lotka-Volterra equations have been used to model not just fluctuations in predator-prey models in population biology but arms races between nations in political science. The widely used Ising models and spin glass models are computational templates, as are the computational rules underlying many agent based models. Important classes of computational templates are the core types of differential equations (together with suitable initial or boundary conditions for solvability) and the standard probability distributions such as the Gaussian, the Poisson, and so on.

## 2 Types of Computational Templates

There are at least four types of computational template. The following classification is based on how on how a piece of formalism comes to be accepted as a computational template.

**Type 1:** The first type arises from a substitution instance of a theoretical template. The substitution of a specific first-order property into the schema is the first step towards a computational template but of course it does not guarantee the computational tractability of the substitution instance. This is in part because the syntactic form of the theoretical template does not determine the syntactic form of the substitution instance. In the Newtonian case, for example, both linear and non-linear equations can result from the substitution of different force functions and the methods needed to arrive at a solution for the non-linear case are usually quite different from the methods used to arrive at a solution for the linear case. Because Type 1 templates are rooted in a subject matter specific theory, even if a very broad theory such as classical mechanics, their origin provides no reason why they should be applicable in other areas and so qua Type 1 templates, our question remains without an answer. Note that the answer cannot rest simply on the fact that classical mechanics is a very general theory covering many material systems, because the point of application of that theory occurs at the level of the computational template and not at the level of the theoretical template, and computational templates vary enormously in their mathematical form.

**Type 2:** The repeated success of a given computational template can result in its elevation to a stylized, abstract, computational template, separated from its original interpreted theoretical context and available to model other, often different, types of phenomena. This is a second recognizable kind of computational template, an off the shelf tractable device that can be opportunistically justified at the system level by analogical reasoning from its previous successful applications to systems that are recognized as being structurally similar. These are the kinds of templates found in various Methods textbooks. These general equation forms – Laplace's equation, Poisson's equation, the diffusion equation, and many other well-known general equation types, or the familiar Gaussian, Poisson, binomial, and other statistical distributions, to mention only two well-known kinds – transcend specific theories and their subject matter. To take Thomas Kuhn's work [6] as a reference point, he recognized that the ability to analogically transfer showpiece successes exemplars as he called them such as the application to a mass on a vibrating spring of a model for a simple harmonic oscillator to new situations such as a simple pendulum was a key part of a scientist's training. But because Kuhn's exemplars are subject-dependent, they lack the subject-transcendent quality of this kind of computational template. As with Type 1 cases, it remains unexplained why the transfer of a piece of formal apparatus from one system to another is successful.

**Type 3:** A third kind of computational template arises from the fact that exactly the same formal 'theoretical' template can sometimes be constructed from radically different assumptions. For an example of this in terms of constructing the diffusion equation, I refer you to [3]. Although each of those constructions rely on assumptions that are motivated by subject matter specific considerations, because exactly the same template is reached from very different assumptions, the resulting template is not tied to the subject matter. It is this fact that explains why the diffusion equation can be applied to both discrete and continuous phenomena, as long as one is willing to accept the empirical correctness of the limit assumptions involved. One answer to our question thus rests on the fact that computational templates can often be constructed from a surprising variety of different starting points, using appropriate idealizations and limit assumptions, a fact that allows us to understand why they are widely applicable. These non-theoretical computational templates sit at the intersection of multiple subject domains, which is one reason they tend to figure prominently in complexity sciences. (For a simple example take the logistic equation, which has been used to model the growth of insect populations and the spread of languages)[1]. Of course, recognizing this fact raises a new explanatory question at a deeper level, which asks why those different starting points fit the systems that they do, but remember that Wigner's puzzlement was not about the applicability of mathematics but about its 'unreasonable' effectiveness. The more we can do in terms of explaining why a given part of mathematics applies to multiple, different, systems, the less unreasonable the success of mathematical modeling will appear to be.

The difference between Type 2 and Type 3 templates is that in the Type 2 case the justification is made at the level of the template itself using analogical inference, whereas the justification for a given Type 3 case is in terms of the construction from more basic elements.

**Type 4:** A fourth kind of computational template arises when a set of very general structural considerations leads, by a single derivation, to a widely applicable equation based template. One example is the Poisson process in statistics, which can be derived from four simple structural assumptions. Another example is contained in the well-known paper of Barabasi and Albert [1] on generating scale free networks. The conditions imposed on such a network is that a) it grows by the addition of new vertices, rather than remaining static and b) that there is a preferential attachment of new nodes to existing well-connected nodes in the sense that the probability for a new node to be attached to an existing node is proportional to the number of nodes already attached to the existing node. The result is that the probability of a node being connected to k other nodes is given by a power law. In contrast, in the Erdos-

---

[1] Sometimes, the same formalism can be arrived at from subject matter specific considerations and also from a more abstract perspective. In an unpublished paper, Tarja Knuuttila and Andrea Loettgers show how Lotka and Volterra separately arrived at essentially the same computational template, the former by analyzing a specific fisheries problem, the latter by analyzing an abstract set of dynamical constraints.

Renyi random graph model, the connections are made randomly rather than preferentially and the probability of connectivity is then given by a Poisson distribution.

Such networks are said to represent a wide variety of systems[2]. One common example is the World Wide Web, with the nodes representing sites and an edge representing a link from one site to another. A second application is to the number of citations for journal articles, where the nodes represent individual articles and a directed edge represents a citation of the in-node by the out-node. The distribution is claimed to follow a power law [7]. The philosophically important point is that this fourth kind of template is not based on explicit subject matter specific theory and these power law networks are constructed on the basis of very minimal features. So any network that originated via a process that satisfied the assumptions will be accurately represented. This feature is at odds with the traditional importance of the axiomatic method as employed in science, which axiomatizes a subject matter specific theory such as the von Neumann-Morgenstern axiomatization of utility functions [8] or the axiomatization of status characteristics theory in sociology [2]. In the case of computational templates, rather than starting with coherent pieces of a subject matter dependent theory, we start with bits and pieces of syntax that represent relevant aspects of the system. This piecemeal construction does not provide evidence for the disunity of science; to the contrary, it re-introduces a level of generality and unification into the representations and as was the case for Type 3 templates, understanding how this generality occurs reduces the sense of unreasonableness at the effectiveness of the template.

## 3   Templates and Solution Methods

'Effectiveness' can apply to a variety of mathematical virtues. In the case of simulations it can, amongst other things, mean effectiveness at representing parts of the world or it can mean being effectively applicable to target systems. The virtues involved are, roughly, representational power and computability in practice. Wigner's question tends to emphasize the former; my interests are primarily in the latter. So, one task in beginning to construct an answer to our question is to consider the sense in which we can separate a template and a solution technique for that template. A formal solution technique is an algorithmic method used to get from a computational template to a mathematically true statement

---

[2] I recognize that there is considerable debate as to how well some of these examples fit the underlying model.

and a factual solution technique is a formal solution technique supplemented with a method for selecting the empirically possible solutions. Like templates, but for a different reason, solution techniques are not themselves representational, although they produce a transition from a template to something that is representational. Well-known examples of solution techniques are optimization procedures on two dimensional surfaces or on an energy landscape, separation of variables for certain kinds of differential equations, and the development of identifiability conditions in econometric models. Some of these techniques are analytic and some are numerical but they must be available in order for the formal object to count as a computational template.

In order to see in more detail how we get from a theoretical template to a computational template, take any differential equation as a representative formal device. In the traditional syntactic account of theories, the application of such an equation with suitable initial or boundary conditions to a system consists simply in the existence of some derivation using the fixed apparatus of deductive logic. In practice, however, a substitution instance of this kind of theoretical template must be supplemented by a *solution method* in order to actually calculate outcomes from the computational template and the methods will vary depending upon the form of the substitution instance of the theoretical template. It is therefore tempting to say that a computational template is a specific substitution instance of the theoretical template that is computationally tractable, augmented by a solution method. In some cases of equation based models this is correct, because the solution methods do indeed form a separate set of techniques that are applied to the model in rather the same way that the deductive apparatus of traditional syntactically formulated theories was taken to be a separable apparatus. So, for example, using a Monte Carlo method for approximately integrating a function is a computational technique that can legitimately be considered a separate component of applying a model. As a second example, the fourth-order Runge-Kutta method for arriving at approximate numerical solutions to ordinary differential equations is a well developed technique that can also be considered separately from the model itself. In both techniques, model specific adjustments may need to be made – importance sampling in the Monte Carlo approach will often be function specific, for example – but the solution method is still a separate component of the model.

In such cases in which the solution technique is separable from the model, an answer to the effectiveness concern is to note that the same solution technique can be applied to a variety of different simulation models, thus providing a unifying explanation for the effectiveness of those models.

In other cases, and this is particularly clear in the case of some agent based models and with cellular automata models, the method itself is an integral part of the computational template and the template is self-contained in terms of including its own solution method. For example, consider a standard fitness landscape model populated with agents equipped with a search algorithm. What counts as a solution in this case? Any future state in the dynamical evolution of this model can be produced by running the model and it will work out its own development using only syntactic resources that are purely internal to the model. There is no global representation to which an external method can be applied. This contrasts with many traditional equation based models in that although analog computers can be used to dynamically produce future states, this cannot be done syntactically unless the model is supplemented with further solution techniques that lie outside the model proper. In this second type of case, we have to fall back on the kind of explanations given for Type 3 and Type 4 computational templates, when they are available.

The framework described above will be brought to bear on issues concerning how agent based simulations differ from simulations that use a higher level conceptualization of the phenomena, how those higher level concepts are connected to emergent phenomena, and in what sense simulations can be considered as numerical experiments.

# References

[1] Barabasi and Albert. Emergence of scaling in random networks. *Science*, 286:509–512, 1999.

[2] P. Humphreys and J. Berger. Theoretical consequences of the status characteristics formulation. *American Journal of Sociology*, 86:953–983, 1981.

[3] Paul Humphreys. Computational science and scientific method. *Minds and Machines*, 5:499–512, 1995.

[4] Paul Humphreys. Computational models. *Philosophy of Science*, 69:S1–S11, 2002.

[5] Paul Humphreys. *Extending Ourselves*. Oxford: Oxford University Press, 2004.

[6] Thomas Kuhn. *The Structure of Scientific Revolutions*. Chicago: University of Chicago Press, 1962.

[7] S. Redner. How popular is your paper? *European Physics Journal B*, 4:131–134, 1998.

[8] Richard von Neumann and Oskar Morgenstern. *The Theory of Games and Economic Behavior*. Princeton: Princeton University Press, 1944.

[9] Eugene Wigner. The unreasonable effectiveness of mathematics in the natural sciences. *Communications in Pure and Applied Mathematics*, 13:1–14, 1960.

# Using the CoSMoS Process to Enhance an Executable Model of Auxin Transport Canalisation

Philip Garnett[1], Susan Stepney[2], Francesca Day, and
Ottoline Leyser[1]

[1] Area 11, Department of Biology, University of York, YO10 5YW, UK
[2] Department of Computer Science, University of York, YO10 5DD, UK

**Abstract.** We describe our use of the CoSMoS process to structure an incremental change of a biological simulation. The domain is auxin transport canalisation. An existing simulator is
refactored to handle aspects of 2D and 3D space more efficiently,
and enhanced to include more realistically-shaped plant cells.
The CoSMoS process supports clear separation of concerns, allowing us to concentrate on the biological model and the implementation decisions separately. This gives a clear and well-
justified simulator design that can be validated by biologists, yet
still allows efficient implementation.

## 1   Introduction

Biological systems present many challenges to science, particularly due to
the complex nature of biology itself. Many biological processes are highly
connected, making it hard to study them in isolation. It is frequently
difficult to get good quantitative data; even good data might lack part
of the larger picture. These factors and many others make it difficult to
form good assumptions about how a biological process is being regulated;
our solutions might reflect our lack of knowledge, rather than offer insight
into the real process.

Increasingly biology is looking to modelling to help progress understanding. Developing a simulation of a biological process is a challenging
task in itself, but doing so can assist with some of the problems. The
modelling process requires the builders to go systematically through the
information and data about a system, ideally with experts in the field.
Simply going through this modelling process can highlight new areas of
focus, or problems and gaps in understanding. The resulting simulation

and models can also be a tool for the generation and testing of hypotheses, hiding some of the complexity of the real system but capturing enough to allow the study of the process of interest.

The level of abstraction in a model is critical. Too high, and we risk ruling out the possibility that simulations will produce interesting emergent behaviours that are observed in the real system. Too low, and the simulations produced could be difficult to work with, understand and validate. These factors make the design decisions made when producing a simulation important, as they determine the balance between these conflicting requirements.

A simulation must be developed using a rigorous process of design, implementation, and validation if it is to be scientifically respectable. Additionally, a useful simulation will need to be upgraded and enhanced in a principled manner as its requirements change to address new research questions. The CoSMoS (Complex Systems Modelling and Simulation) process [1] provides a flexible approach designed to support the modelling and analysis of complex systems, including the design and validation of appropriate computer simulations.

We have previously used the CoSMoS process to guide the initial development of a simulation of an abstract tissue level model of plant cells [14]. Here we present work using the same process to guide modification and enhancement of this existing system, by improving the model of space, and allowing more naturally shaped cells. This work helps demonstrate how the CoSMoS process can be used in an incremental manner.

In §2.1 we overview the CoSMoS process as used for modelling, designing, and implementing biological system simulations. In §2.2 we discuss the use of UML as a suitable modelling language to support this process. In §2.3 we give an overview of the initial auxin model. We then use the CoSMoS process components to structure the remaining sections. In §3 we introduce the Research Context. In §4 we summarise the biological Domain Model. In §5 we discuss the issues relating to modelling space that we are addressing in this increment. In §6 we discuss how the Platform Model has been updated using the CoSMoS process. In §7 we conclude with a discussion of our experiences.

## 2    Background

### 2.1    CoSMoS Process: The modelling lifecycle

Described in detail by Andrews et al. [1], and used in our earlier work [14], the CoSMoS process provides a systematic approach to building models and simulations of complex systems, including the biological system of

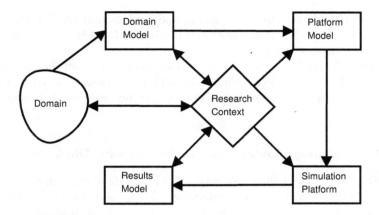

**Fig. 1.** The components of the CoSMoS process [1, fig.2.1]. Arrows indicate the main information flows during the development of the different components. There is no prescribed route through the process.

interest here. The CoSMoS process does not include a defined end point: the process is incremental, aimed at supporting a series of simulations. We [14] and others [34] have successfully used this process to assist in the production of simulations of complex biological systems. Summarised in figure 1, the process contains the following components (summarised from [1, 14]):

**Research Context** : the overall scientific research context. This includes the motivation for doing the research, the questions to be addressed, and the requirements for success.

**Domain Model** : conceptual "top-down" model of the real world system to be simulated. The domain model is developed in conjunction with the domain experts, with its scope determined by the Research Context. The model may explicitly include various emergent properties of the system.

**Platform Model** : (called the Software Model in [14]) a "bottom up" model of how the real world system is to be cast into a simulation. This includes: the system boundary, what parts of the Domain Model are being simulated; simplifying assumptions or abstractions; assumptions made due to lack of information from the domain experts; removal of emergent properties (properties that should be consequences of the simulation, rather than explicitly implemented in it).

**Simulation Platform** : the executable implementation. The development of the Simulator from the Platform Model is a standard software engineering process.

**Results Model** : a "top down" conceptual model of the simulated world. This model is compared with the Domain Model in order to test various hypotheses. (This part of our research is beyond the scope of this paper.)

## 2.2    Modelling biology and simulations with UML

UML (Unified Modelling Language) [27] is a suite of diagramming notations designed to aid in the development of large object-oriented software engineering projects by groups of developers working in teams.

Although UML is normally used in conjunction with an object-oriented programming language, it is well suited to agent-based modelling [26], where an agent can be thought of as an object with its own thread of control, allowing highly parallel systems of multiple agents. Biological 'agents', such as cells and proteins, can be modelled as UML agents. This relatively natural mapping between biological agents and their UML counterparts means that much of the structure of a biological simulation can be well-represented by UML. There are a number of published cases where UML has been successfully used to assist the production of biological models [10, 14, 16, 34, 44].

We have found that UML diagrams (in conjunction with traditional biological 'cartoons') are relatively accessible to biologists, allowing these domain experts to provide input to the model of the simulation without the need to understand the implementation details.

## 2.3    Auxin transport canalisation model

Auxin was one of the first plant hormones to be discovered, 130 years ago by Charles and Francis Darwin [9]. Understanding auxin's functions still presents many challenges to plant science as it is involved in diverse aspects of plant patterning and development. Computational modelling plays an important role in auxin transport research [13].

We are using a UML-based approach within the CoSMoS process to design and build a simulation of auxin transport canalisation in the plant *Arabidopsis*. Our initial model and simulation is described in [14]. This approach allows us to build models containing the biological objects that we believe to be involved in auxin canalisation, and then produce simulations to test various hypotheses about the biological processes of interest. If an hypothesis is correct we should see the correct emergent behaviour when the simulation is run; if not we can then return to the UML models and implement our next hypothesis. If all our hypotheses fail to produce the emergent behaviour of interest we might have to return to a different part of the process.

In this paper we describe an enhancement to our initial model in [14]. The most significant modifications are in the Platform Model and associated Simulation Platform. We revisit significant assumptions about what should be removed from the domain model. The main progress made in the design and implementation of our models has been with the handling of the simulation space, allowing the cells of the tissue modelled to be more naturally shaped. The improved 2D simulator has been adapted into 3D.

## 3   The Research Context

The auxin transport community studies many different aspects of auxin transport. These include, but are not limited to: auxin transport canalisation [37, 38]; shoot branching regulation [21, 22, 28]; leaf venation [41]; and phyllotactic patterning [20, 35]. These processes are concerned with the developmental patterning of a plant, at both the tissue level and that of the whole plant.

Our research sits within this wider community; it uses background biology derived from the literature, and from wet-lab experiments carried out in the Leyser group (for more information see §4). We primarily focus on modelling the process of auxin transport canalisation, within the context of shoot branching regulation.

There are many published mathematical models of auxin transport. We have chosen to develop executable models as we believe this modelling technique lends itself to biological systems, and can offer an alternative perspective [11, 13], particularly as we are modelling the PIN protein transporters at a reasonable level of detail.

Our models focus on the question of PIN cycling and its role in canalisation, and we aim to test different regulatory mechanisms of PIN cycling.

## 4   The Domain Model: auxin transport canalisation

The domain of our model remains auxin transport. §4.1 summarises the background biology used as input to the Domain Model. The full model is informed by more detailed biological information than is summarised here; we direct interested readers to reviews of auxin transport [4, 3, 7]. §4.2 summarises the way this biology is captured in UML diagrams.

## 4.1   The Biological Domain

We are developing models to investigate the formation of auxin transport
canals in plant tissues. This process of canalisation and its regulation are
not fully understood.

Canalisation can be thought of as a self-organising process, where
auxin in cells promotes its own transport between cells through the tissue
of the plant [37]. In canalisation the transport goes from a source, an
area where auxin is accumulating, to a sink elsewhere in the tissue. The
link that forms between these two sites is called an auxin canal, and
the process by which it forms is canalisation [23, 24, 39]. The transport
of auxin between cells is dependent on membrane localised transport
proteins, of which the ABCB and PIN transporters are two prominent
families [2, 12, 15, 31, 43, 45]. We are primarily interested in the PIN
family of transporters. PIN proteins are often distributed asymmetrically
around the membrane of a cell. This asymmetry enables directional auxin
transport, which is central to canalisation.

We are particularly interested in canalisation within the context of
shoot branching regulation. Shoot branching is the process where lateral
axillary buds on the main stem of a plant activate and grow into branches
[22]. Auxin produced higher up the plant inhibits the growth of lateral
axillary buds, a phenomena known as apical dominance [8]. If the auxin
sources inhibiting a bud are removed by decapitating the plant the bud
is released and is able to grow. This can be reversed by application of
auxin directly to the site of decapitation. We believe that the bud is able
to grow only when it can export its auxin into the main stem.

The vascular link between an active growing bud and the main vascular tissue in the stem requires auxin transport canalisation from the bud
to the stem to trigger its differentiation. It is this canalisation process
we would ultimately like to model, as understanding canalisation at this
position in the plant could help with the understanding of shoot branching. In order for an auxin transport canal to form between the bud and
the main stem, the stem vasculature must behave as a relatively strong
sink when compared with the surrounding tissue. If the stem vasculature is already transporting large amounts of auxin from higher up the
plant, its sink strength is reduced, the canal does not form, and the bud
does not activate and is unable grow into branch. However, if the level
of auxin in the stem starts to fall, its sink strength increases; this allows
canalisation to occur and a canal is able to form. Auxin can be exported
out of the bud, causing it to activate and grow into a new branch.

This process of bud activation has been successfully modelled mathematically on a tissue and whole plant scale [32]. However, there are
processes occurring at the cellular level that are not fully understood.

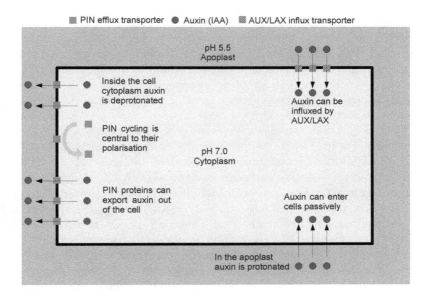

**Fig. 2.** Domain Model PIN Localisation: Auxin transport into and out of cells is central to canalisation. Protonated auxin in the apoplast is able to enter the cell passively, or to be actively influxed by AUX/LAX transporters. Once inside the cell the majority of auxin is deprotonated and is therefore unable to leave the cell unaided. This is often known as the Acid Trap hypotheses [36, 33]. PIN transports are important to the efflux of auxin from cells. The regulated cycling of PINs on and off the cell membrane causes them to become localised asymmetrically around the cell membrane. This process is not fully understood, but is critical to the directional transport of auxin in tissues, and the process of canalisation.

Auxin has an interesting cell biology that is responsible for some aspects of its behaviour (figure 2). Auxin is a weak acid and therefore some auxin is able to enter cells passively from the more acidic apoplast (intercellular space) by crossing the cell membrane. It can also be actively transported into cells by AUX/LAX influx carrier proteins [30]. Once in the pH-neutral cytoplasm the majority of auxin is deprotonated, and therefore unable to recross the membrane passively. It is essentially trapped, a phenomena known as the Acid Trap hypothesis [33, 36]. Auxin is only able to leave the cell via efflux transport proteins. We are interested in the PIN family of transporters, as they are found to be polarly localised in cells that form auxin transport canals and are therefore very likely to be central to canalisation [40].

The process of canalisation has been the focus of much prior work over a long period. Sachs suggested a model where auxin facilitates its own flow: both the ability of a cell to transport auxin and the polarity of the auxin flow increase with the amount of auxin being transported [39]. Therefore as the transport capacity increases the cells in the canal become better sinks and draw in more auxin from their neighbours. Mitchison modelled this process mathematically and was able to show it to work [23, 25]. Mitchison's models predict canals of high flow and low concentration, where as experimental evidence suggests that there is both high flux and *high* concentration [5, 42]. Kramer later produced models that showed that the addition of the AUX/LAX auxin influx proteins can allow for canals of both high flux and high concentration [17].

We now have more information about the biology of canalisation. Experiments show that auxin is up-regulating its own transport by increasing the amount of PIN protein available to transport auxin [29]. Thus the more auxin in a cell, the more it can transport. This has been further confirmed by experiments showing that if the negative regulators of PIN accumulation are removed, auxin transport increases and the stem is able to transport more auxin [6]. The other key part of the process is the localisation of PIN to provide the directional transport of auxin. However, the mechanism of PIN localisation is not understood.

PIN proteins are therefore of great interest to the canalisation process as they export auxin out of the cells, and their polar localisation patterns are responsible for complex transport patterns in a number of plant tissues [18, 19]. However, what directs the PIN in the cells into the observed polar patterns remains an important question: if PIN is positioned by detection of auxin flux, as Sachs suggests [39], what is it in cells that is detecting auxin flux? This is one problem our simulations aim to address.

## 4.2   Domain Model UML

The UML used to capture the Domain Model has not changed significantly during the development process. We briefly summarise it here, but direct interested readers to our previous paper for more detailed discussion [14].

**Domain Model use cases.** These capture a high level view of what the system does, such as the regulation of proteins and hormones.

**Domain Model class diagram.** This captures the biological entities of interest as objects and classes. Objects map naturally to biological

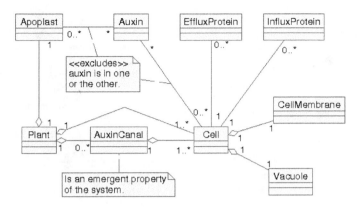

**Fig. 3.** Domain Model class diagram [14, fig.4].

entities such as proteins, hormones, and cells. Cells themselves are composed of a number of objects such as membranes, cytoplasm and vacuoles, which are associated with each other in space. We also need to regulate the production of agents like proteins and hormones, which is done by cells. See figure 3.

**Domain Model state diagrams.** These are among the most useful of the Domain Model diagrams for communicating with the Domain Experts, as they appear to map well to the way these biological processes are understood. State diagrams capture how an object changes through time. They are able to show the different possible states of the biological objects, and how an object moves from one state to another. Some spatial information can also be captured by state diagrams, as the changes can be associated with a location, within and outside a cell. For example the possible state changes that the auxin object can undergo are different depending on whether it is inside or outside a cell. State diagrams map neatly to the traditional biological 'cartoon' showing process occurring in cells (such as figure 2). The behaviour of auxin can be cross-referenced between the 'cartoon' and the Domain Model auxin state diagram (figure 4).

## 5   Modelling Space

Here we discuss an important part of the model that was not explicitly dealt with in the first increment [14]: space.

The simulation space is part of the biological domain that cannot easily be captured using UML, and might be based on assumptions that

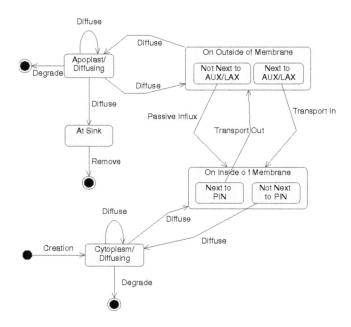

**Fig. 4.** Domain Model auxin state diagram [14, fig.6].

could escape recording. The space in which our biological entities exist is implied in the UML. We can see from the domain class diagram (figure 3) that we are representing part of a Plant built from a number of Cells (each with a CellMembrane and Vacuole), surrounded by Apoplast. However the nature of the space is not captured, nor is any information about how the objects such as CellMembranes or Vacuoles are arranged into Cells, nor how the Cells and Apoplast are arranged into a plant tissue. This information might seem obvious, since it is easy to imagine (particularly if you work in the field of plant science) what a small 2D section of plant tissue might look like. This aspect is easy to capture with a more traditional 'cartoon' and explanatory documentation.

In our initial simulation the assumption is made that a 2D rectangular 'box' is an adequate representation for a plant cell. Therefore the initial simulation is limited to 2D cells of four straight sides. This is a reasonable simplification to make; mature cells in the stem of a plant are often fairly block-like in shape. However, auxin transport canals also form through tissues with cells of varying size and shape, particularly at the interface of a bud and existing vascular tissue. Therefore being able to test the behaviour of our hypothesised regulation of PIN localisation in cells of more natural shapes would be interesting both from a biological and simulation point of view.

Linked to this is the need to try to investigate the effect that 3D cells would have on the behaviour of the hypotheses. There are a number of differences between real 3D cells and simulated 2D cells that might have an effect on the PIN localisation. Being able to simulate even a small number of 3D cells could provide interesting insight into the effect of abstracting 3D cells into 2D. Early simulations have been done in 3D, but it is not well implemented in the initial simulation. We also want to allow for more naturally shaped 3D cells. The first of these issues are linked to the way in which space (the environment of the agents) in the model is handled. This impacts a number of key areas: the interaction between the agents and the space, and how the space is split up into cells and the other structures in the plant tissue.

These modifications are more about changes in the level of abstraction assumed during the development of the Platform Model, about how the simulation is to be constructed from the Domain Model. Sometimes it is possible to change existing simulation code to allow for the change in abstraction. In our case the changes are significant, and the development process of the first simulation highlights a number of areas where improvements could be made.

# 6   Platform Model

The Platform Model includes all the extra components that allow the simulation to run. This includes all the processes required to get the simulation to a point where it is able to start, such as generating the space and populating it with cells.

The Platform Model has three kinds of information: biological processes captured directly from the Domain Model; biological processes required for the proper functioning of the simulation, but not of explicit interest to the Research Context, implemented with regard to efficiency rather than biological fidelity; instrumentation and other such aspects of a simulation that are not part of the Domain, but are needed to observe and document the simulation results.

Throughout the continued developmental process it is the Platform Model that has seen the most change. Not only have we made efforts to make the simulated space more realistic with respect to the real plant, but huge improvements have been made in the data output from the simulations and the organisation of the code.

### 6.1 Platform Model UML

**Platform Model use cases**: these capture the user requirements for using the simulator, the traditional use for use cases in software engineering. These are unchanged from the original version [14].

**Platform Model class diagram.** This is produced from the Domain Model class diagram, with all emergent properties (such as the Auxin Canal) removed. This high level diagram shows mainly the biologically relevant parts of the model, and is unchanged in this iteration (figure 5).

**Platform Model class diagram, implementation level.** As we move towards code, implementation level data structures are added to the class diagram. §6.2 discusses the changes to the implementation level Platform Model class diagram.

**Platform Model state diagrams.** These follow the Domain Model state diagrams and remain largely unchanged from the original version [14].

As the simulator increases in complexity, keeping the high level and implementation level Platform Models distinct becomes increasingly important. Things that are not biologically relevant, but are needed in a simulator, such as the ability to easily checkpoint to allow restarting, add complexity to the model that biologists do not need to see. We therefore omit such detail from the high level Platform Model diagrams discussed with the biologists, and retain it in implementation level Platform Model diagrams used by the developer.

### 6.2 The Division of Space

The main changes we made in moving from the initial to the enhanced version were to the way the space is handled in the Platform Model and simulation.

The initial version treats the space as a largely homogeneous area, a grid of pixels, on which cell membranes and vacuoles are drawn, dividing the space into separate areas. Some areas are associated with objects like Vacuole and CellMembrane; other areas are essentially null.

A CellMembrane is a continuous line enclosing the cell (figure 6A). It is straightforward to define a cell membrane if it is built from straight line segments. However it is more difficult to define realistic-shaped cells with curved membranes (figure 6B) using this approach. The membranes would need to be drawn correctly somehow, and then read into the simulation. It would be easier to place the cells into the space as continuous

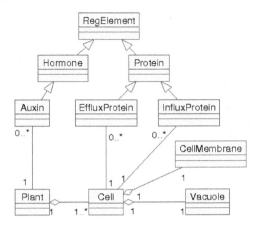

**Fig. 5.** Platform Model class diagram [14, fig.10]. Note that space is not explicitly dealt with, rather it is generic unless something like a CellMembrane object is put into a position.

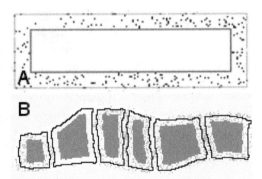

**Fig. 6.** (A): Section of visual output from the initial simulator. The thin line of the cell membrane (outer grey line) is drawn into the space to define the cell. The vacuole is defined by drawing another membrane (darker grey line). This is a simple task for boxes, but more difficult for natural shapes. (B): Section of visual output from the enhanced simulator, showing a continuous curved membrane (black line).

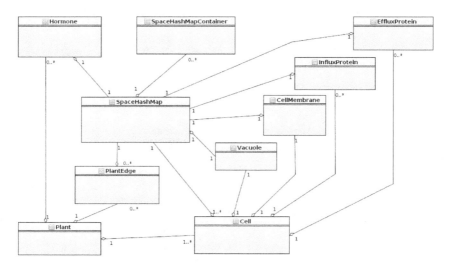

**Fig. 7.** Implementation level Platform Model class diagram of the initial simulator. All objects in space require access to a singleton class SpaceHashMap that provides them with information about the space they are in, via the Space-HashMapContainer class. As more kinds of space are needed, the resulting code becomes inefficient and untidy. (Inheritance has been left off this diagram to improve readability.)

areas of cytoplasm, and then determine the position of the membranes around the edge (which is how it is implemented in the enhanced version). A new method of handling the space needs to be able to address such issues. We also want it to be easier to extend the range of different types of space that could exist in the simulation.

In the initial version of the model, all space is described by a single object. Figure 7 shows the relevant part of the implementation level Platform Model class diagram. A single class, SpaceHashMapContainer, has different attributes that allow it to represent all of the different types of space in the simulation, depending on the values the attributes are given. However, the complexity and size of this class increases each time we add a new kind of area of space in the simulation.

Another significant issue with having all the kinds of space specified in a single class is that some of the methods in the class need to behave differently depending on what the kind of space is. This increases the complexity of the individual methods in the class. The organisation of the code also suffers from having added the space to the model, rather than it having been designed with space in mind.

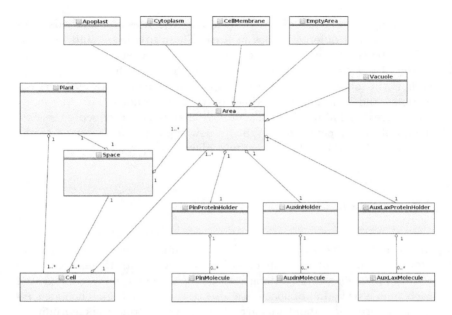

**Fig. 8.** Implementation level Platform Model class diagram of the enhanced simulator. The space is now built from different child classes of the Area class, each with a holder looking after the different Molecules. The Space contains many areas which compose a single Plant, the Plant has many Cells. Cell requires access to the Space directly but also contain within it a list of all its associated Areas. A Cell does not directly contain any Molecules. (Inheritance of the different Holder and Molecule classes are not shown, to improve diagram readability.)

For the enhanced version, we refactor the code to handle the space in a more area-specific manner, to improve its structure and extensibility, and to allow more natural-shaped cells.

In the initial model, space is general unless it is given a particular type. In the enhanced model, all the space is given an area type. An abstract class Area has attributes common to all the different types of area in the simulation. Sub-classes extend the abstract Area class into more specific kinds of space. Currently there are five types of area. Cells have Cytoplasm, Membrane, and Vacuole. Outside the cells there is Apoplast: the cell walls. Finally there is EmptySpace; this is used to allow more elaborate shapes of space to be used in the models, and is not processed. Apoplast areas separate all the cells from each other, and also separate cells from EmptySpace. See figure 8.

The abstract **Area** class contains many attributes and methods common to all the different types of area. These attributes and methods tend to be the system aspects of the class, such as accessing the colour of the object or its position in the space. The specific area type then adds extra methods that give that space more biologically specific behaviour, and if necessary overload particular methods. This has many advantages, including simplicity of code maintenance reducing the likelihood of introducing errors. When a new type of space is added to the model much of the code is already in place.

## 6.3 Agents in Space

In the initial version, the code that determines how the agents move around in the simulation space is held in the agents themselves. This results in the classes describing the agents becoming more complicated each time a new kind of space is added to the simulation. The agent requests information about its current environment from the environment directly. It then uses this to make an appropriate decision about what it would do. There is also an inconsistency in where the agents are stored. Figure 7 shows that auxin (**Hormone** objects) are held in the **Plant** class, but the proteins are in the **Cell** class. This makes *biological* sense, since the PIN and AUX/LAX proteins do not leave the cell, but auxin does. However it makes better *implementation* sense to think as all three as being held in the **Space**, and whether or not this is in a **Cell** is determined by what the space is. This is the case for the enhanced simulator, as shown in figure 8.

The movement of agents is also the responsibility of the **Space** in the enhanced simulator. Each **Area** sub-class that can have agents contains an **AgentHolder** with methods for storing the agents that are contained within it. The different **AgentHolder** sub-classes (such as **AuxinHolder**) for each agent inherit properties from the parent **AgentHolder**, but are also given specific behaviours. The **AgentHolder** classes accept incoming agents to their area. The movement of the agents is controlled by the **Area** sub-class, which has methods for moving any agents in the relevant **AgentHolder**. This puts the responsibility of moving agents onto the **Area** class. Therefore when a new kind of space is added, the areas are updated to allow agents to move into this new kind of space. These changes are reflected in figure 8 (the inheritance from the abstract **AgentHolder** class and the **Molecule** class are not shown, to improve the readability of the diagram.)

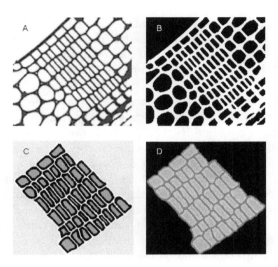

**Fig. 9.** Processing sections of plants into the model. If the section photos are of high enough quality the processing can be done automatically. (A) Photographic section from a real plant, tided up to allow it to be processed. (B) Image processed for reading into the model: black areas will become Cytoplasm, white areas Apoplast. (C) Image modified by hand to isolate a patch of cells: light grey areas will become EmptySpace. Vacuole areas are then added automatically (dark grey). (D) Template as it finally appears in the simulation visualisation. CellMembrane areas are added automatically at the interface between Cytoplasm (here light grey) and Apoplast (here dark grey).

## 6.4 Space from Templates

The more natural-shaped cells are defined using templates derived from images of real plants.

Figure 9 shows the lifecycle of a template: it starts as an image of a section of a plant, and ends as a representation of the simulation space. Templates can either be generated automatically (normally with a little manual processing), or fully by hand. They need to contain only three pieces of information: the areas of the space that are empty (not active as simulation space but required to be spatially present); the areas that are apoplast; and the areas that are cells. The template is then processed to add vacuoles into the cells. These are not added directly from the image being used, because simulated 2D cells need smaller vacuoles than are shown in sections of real 3D cells. Instead they are added automatically by filling the centre of the cell a certain amount (see §6.5 for discussion of this). Cell membranes are also added automatically around the cytoplasm. Once the vacuoles and cell membranes have been added into the space we essentially have areas presenting cell cytoplasm, cell membranes, vacuoles, apoplast and any empty areas. All are displayed as different colours in the image (shown as different shades of grey in the figure).

In the simulation the space is created to match the pixel size of the template, and the entire space starts off as apoplast. Each pixel of the template is then read and its colour determines what it is in the space. The next task is to group areas of continuous cytoplasm and the vacuole inside them into the more abstract notion of a cell. In a plant, a cell is essentially a container of elements that need to be held together. The elements have no concept of togetherness, they are just associated in space. The way the different elements interact is through the common environment. In the simulation a cell is more abstract. It is similar in that it contains lists of all of its spatial contents but it also needs methods to create more proteins or hormones when they are required. Essentially the nucleus of a real cell, which regulates what is expressed, is part of the more abstract Cell class in the simulation. The Cell class provides access to the common environment, to allow cell regulation.

## 6.5   3D Space

Our initial simulator version can handle 3D models, but not very efficiently. The enhanced simulator space is implemented by ensuring that all Areas know who their neighbours are, and therefore the move to 3D is much simpler as it mainly involves giving the Areas more neighbours. The code for the 2D and 3D versions of the simulator are therefore very similar, which makes it much easier to maintain.

We can either generate block-shaped 3D cells from algorithms, or naturally shaped cells by stacking prepared 2D templates together in a careful order to create a 3D space. This requires three kinds of templates, containing: only Apoplast; Apoplast and Cytoplasm; Apoplast, Cytoplasm and Vacuole.

We are interested in 3D simulations to investigate how our hypotheses behave in 3D, and the effect of using 2D simulation, particularly on the effects of vacuoles. Compare the possible paths an auxin molecule can take in a 3D cell with a large vacuole to that of a 2D cell with a large vacuole. We can see from figure 10 that in a 3D cell taking the path through the vertical section is much longer than taking a path through the horizontal section at roughly the position of the dashed line. In the 2D cell there is only the vertical path. All other diffusing agents will have the same problem. This could have an effect on auxin transport in a 2D

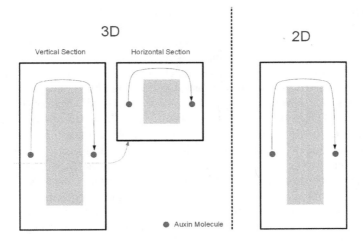

**Fig. 10.** Comparison of possible paths of auxin molecules (or other agents) in 2D or 3D cells. In the 3D cell the auxin has the possibility of taking a short path to the same position. This is not possible in a 2D cell with only one path.

tissue. We can use the 3D simulation to help calibrate the required size of the 2D vacuole.

## 7 Discussion

We have used the CoSMoS process to produce an incremental change to a pre-existing CoSMoS-based model and simulator. The enhanced simulator has improved performance, allowing us to run simulations of canalisation over larger arrays of cells, and over more naturally-shaped cells. Canals still form in the latter case, indicating that the observed process is not an artefact of the rectangular cells. The biologically-relevant results from this enhanced simulator version will be presented elsewhere; here we discuss the impact of the CoSMoS process on the development.

Continuing to develop our simulations with the CoSMoS process assisted by UML has allowed us to progress in an efficient and systematic way. Using this approach helps us to identify which of the assumptions we made when making the transition to the Platform Model from the Domain Model might need to be reassessed. Both the CoSMoS process and UML have allowed us to see how progressing down a particular development path was increasing the gap between the biology we were trying simulate and how we were implementing it.

The CoSMoS process ensures that at each stage of modelling and simulator development effort is made to understand and acknowledge what decisions have been made and why. It is also flexible enough to work with software engineering tools like UML. UML is able to produce detailed information about the structure of a biological system. It is then possible to extend these UML descriptions of the biology into code skeletons of a simulator, even though the final UML and code include much more than just the underlying biology. That underlying structure should be visible (visibility can be improved by maintaining a separate Platform Model and Refined Platform Model), and areas where it has had to change or has been deliberately changed (such as the removal of emergent properties) can be highlighted and the reasons made clear. UML diagrams, particularly state diagrams, can be compared with more traditional biological 'cartoons' to enhance cross-disciplinary communication of model structure and included biology. This can help increase information flow between modellers and domain experts.

Going through the CoSMoS process has allowed us to see that we needed to return to the Platform Model of our simulator to include more natural cell shapes derived from the biology. Both the CoSMoS process and UML allowed us to identify parts of the simulator code that were becoming over complicated and could be improved. From this we were able to improve how the biology of the Domain Model is captured in the Platform Model, and simultaneously improve the simulator code itself.

In the future we may wish to include more aspects of the Domain in the Models and simulation. One important example is growth. Introducing growth into the current simulation architecture would be very difficult to do. Therefore the CoSMoS process could be used to make the transition between the current simulator to a new one in a way that allows us to fully understand the differences between the two simulators produced.

### Acknowledgements

This work is supported by a BBSRC/Microsoft Research CASE studentship and an EPSRC TRANSIT project (EP/F032749/1) summer studentship. Thanks to Lauren Shipley for comments on an earlier draft of the paper, and to the anonymous referees whose suggestions have helped improve the clarity of this final version.

# References

[1] Paul S. Andrews, Fiona A. C. Polack, Adam T. Sampson, Susan Stepney, and Jon Timmis. The CoSMoS process version 0.1: A process for the

modelling and simulation of complex systems. Technical Report YCS-2010-453, Department of Computer Science, University of York, 2010.

[2] A. Bandyopadhyay, J. J. Blakeslee, O. R. Lee, J. Mravec, M. Sauer, B. Titapiwatanakun, S. N. Makam, R. Bouchard, M. Geisler, E. Martinoia, J. Friml, W. A. Peer, and A. S. Murphy. Interactions of PIN and PGP auxin transport mechanisms. *Biochem. Soc. Trans.*, 35(Pt 1):137–141, 2007.

[3] René Benjamins, Nenad Malenica, and Christian Luschnig. Regulating the regulator: the control of auxin transport. *Bioessays*, 27(12):1246–1255, 2005.

[4] René Benjamins and Ben Scheres. Auxin: the looping star in plant development. *Ann. Rev. Plant Biol.*, 59:443–465, 2008.

[5] Eva Benková, Marta Michniewicz, Michael Sauer, Thomas Teichmann, Daniela Seifertová, Gerd Jürgens, and Jiří Friml. Local, efflux-dependent auxin gradients as a common module for plant organ formation. *Cell*, 115(5):591–602, 2003.

[6] Tom Bennett, Tobias Sieberer, Barbara Willett, Jon Booker, Christian Luschnig, and Ottoline Leyser. The arabidopsis MAX pathway controls shoot branching by regulating auxin transport. *Curr. Biol.*, 16(6):553–563, 2006.

[7] Yohann Boutté, Yoshihisa Ikeda, and Markus Grebe. Mechanisms of auxin-dependent cell and tissue polarity. *Curr. Opin. Plant Biol.*, 10(6):616–623, 2007.

[8] Morris Cline. Apical dominance. *The Botanical Review*, 57(4):318–358, 1991.

[9] Charles Darwin and Francis Darwin. *The Power of Movement in Plants.* John Murray, 1880.

[10] S. Efroni, D. Harel, and I. R. Cohen. Toward rigorous comprehension of biological complexity: modeling, execution, and visualization of thymic T-cell maturation. *Genome Res.*, 13(11):2485–97, 2003.

[11] Jasmin Fisher and Thomas A. Henzinger. Executable cell biology. *Nature Biotechnology*, 25(11):1239–1249, 2007.

[12] Jiří Friml, Anne Vieten, Michael Sauer, Dolf Weijers, Heinz Schwarz, Thorsten Hamann, Remko Offringa, and Gerd Jürgens. Efflux-dependent auxin gradients establish the apical-basal axis of arabidopsis. *Nature*, 426(6963):147–153, 2003.

[13] Philip Garnett, Arno Steinacher, Susan Stepney, Richard Clayton, and Ottoline Leyser. Computer simulation: the imaginary friend of auxin transport biology. *BioEssays*, 32(9), September 2010.

[14] Philip Garnett, Susan Stepney, and Ottoline Leyser. Towards an executable model of auxin transport canalisation. In *CoSMoS 2008, York, UK, September 2008*, pages 63–91. Luniver Press, 2008.

[15] Markus Geisler and Angus S. Murphy. The ABC of auxin transport: the role of p-glycoproteins in plant development. *FEBS Lett.*, 580(4):1094–1102, 2006.

[16] Na'aman Kam, Irun R. Cohen, and David Harel. The immune system as a reactive system: Modeling T cell activation with statecharts. In *HCC '01*, page 15. IEEE, 2001.

[17] Eric M. Kramer. PIN and AUX/LAX proteins: their role in auxin accumulation. *Trends Plant Sci.*, 9(12):578–582, 2004.

[18] Eric M. Kramer. Computer models of auxin transport: a review and commentary. *J. Exp. Bot.*, 59(1):45–53, 2008.

[19] Pawel Krupinski and Henrik Jnsson. Modeling auxin-regulated development. *Cold Spring Harb. Perspect. Biol.*, 2(2):a001560, 2010.

[20] Cris Kuhlemeier. Phyllotaxis. *Trends in Plant Science*, 12(4):143–150, 2007.

[21] Ottoline Leyser. The fall and rise of apical dominance. *Curr. Opin. Genet. Dev.*, 15(4):468–471, 2005.

[22] Ottoline Leyser. The control of shoot branching: an example of plant information processing. *Plant Cell Environ.*, 32(6):694–703, 2009.

[23] G. Mitchison. A model for vein formation in higher plants. *Pro. Roy. Soc. Lond. B.*, 207:79–109, 1980.

[24] G. Mitchison. The polar transport of auxin and vein patterns in plants. *Phil. Trans. Roy. Soc. Lond.*, 295:461–471, 1981.

[25] G. J. Mitchison. The effect of intracellular geometry on auxin transport II. Geotropism in shoots. *Proc. Roy. Soc. Lon. B.*, 214(1194):69–83, 1981.

[26] J. Odell, H. Parunak, and B. Bauer. Extending UML for agents. In *Proc. AOIS Worshop at AAAI, Austin, 2000*, pages 3–17, 2000.

[27] OMG. Maintainer of the UML standards. *http://www.omg.org*, 2008.

[28] Veronica Ongaro and Ottoline Leyser. Hormonal control of shoot branching. *J. Exp. Bot.*, 59(1):67–74, 2008.

[29] Tomasz Paciorek, Eva Zazímalová, Nadia Ruthardt, Jan Petrásek, York-Dieter Stierhof, Jürgen Kleine-Vehn, David A. Morris, Neil Emans, Gerd Jürgens, Niko Geldner, and Jiří Friml. Auxin inhibits endocytosis and promotes its own efflux from cells. *Nature*, 435(7046):1251–1256, 2005.

[30] Geraint Parry, Alan Marchant, Sean May, Ranjan Swarup, Kamal Swarup, Nick James, Neil Graham, Trudie Allen, Tony Martucci, Antony Yemm, Richard Napier, Ken Manning, Graham King, and Malcolm Bennett. Quick on the uptake: Characterization of a family of plant auxin influx carriers. *Journal of Plant Growth Regulation*, 20(3):217–225, 2001.

[31] Jan Petrášek, Jozef Mravec, Rodolphe Bouchard, Joshua J. Blakeslee, Melinda Abas, Daniela Seifertová, Justyna Wiśniewska, Zerihun Tadele, Martin Kubeš, Milada Čovanová, Pankaj Dhonukshe, Petr Skůpa, Eva Benková, Lucie Perry, Pavel Křeček, Ok Ran Lee, Gerald R. Fink, Markus Geisler, Angus S. Murphy, Christian Luschnig, Eva Zažímalová, and Jiří Friml. PIN proteins perform a rate-limiting function in cellular auxin efflux. *Science*, 312(5775):914–918, 2006.

[32] Przemyslaw Prusinkiewicz, Scott Crawford, Richard S. Smith, Karin Ljung, Tom Bennett, Veronica Ongaro, and Ottoline Leyser. Control of bud activation by an auxin transport switch. *Proc. Natl. Acad. Sci. USA*, 106(41):17431–36, 2009.

[33] J. A. Raven. Transport of indoleacetic acid in plant cells in relation to pH and electrical potential gradients, and its significance for polar IAA transport. *New Phytologist*, 74(2):163–172, 1975.

[34] Mark Read, Jon Timmis, Paul S. Andrews, and Vipin Kumar. A domain model of experimental autoimmune encephalomyelitis. In *CoSMoS 2009, York, UK, August 2009*, pages 9–44. Luniver Press, 2009.

[35] Didier Reinhardt, Eva-Rachele Pesce, Pia Stieger, Therese Mandel, Kurt Baltensperger, Malcolm Bennett, Jan Traas, Jiří Friml, and Cris Kuhlemeier. Regulation of phyllotaxis by polar auxin transport. *Nature*, 426(6964):255–260, 2003.

[36] P. H. Rubery and A. R. Sheldrake. Carrier-mediated auxin transport. *Planta*, 118(2):101–121, 1974.

[37] T. Sachs. Polarity and the induction of organized vascular tissues. *Ann. Bot.*, 33(2):263–275, 1969.

[38] T. Sachs. Patterned differentiation in plants. *Differentiation*, 11(1-3):65–73, 1978.

[39] T. Sachs. The control of the patterned differentiation of vascular tissues. *Adv. Bot. Res. inc Adv. Plant Path.*, 9:151–262, 1981.

[40] Michael Sauer, Jozef Balla, Christian Luschnig, Justyna Wiśniewska, Vilém Reinöhl, Jiří Friml, and Eva Benková. Canalization of auxin flow by Aux/IAA-ARF-dependent feedback regulation of PIN polarity. *Genes Dev.*, 20(20):2902–2911, 2006.

[41] Enrico Scarpella, Danielle Marcos, Jiří Friml, and Thomas Berleth. Control of leaf vascular patterning by polar auxin transport. *Genes Dev.*, 20(8):1015–1027, 2006.

[42] C. Uggla, T. Moritz, G. Sandberg, and B. Sundberg. Auxin as a positional signal in pattern formation in plants. *Proc. Natl. Acad. Sci. USA*, 93(17):9282–9286, 1996.

[43] Anne Vieten, Michael Sauer, Philip B. Brewer, and Jiří Friml. Molecular and cellular aspects of auxin-transport-mediated development. *Trends Plant Sci.*, 12(4):160–168, 2007.

[44] K. Webb and T. White. UML as a cell and biochemistry modeling language. *BioSystems*, 80:283–302, 2005.

[45] Justyna Wiśniewska, Jian Xu, Daniela Seifertová, Philip B. Brewer, Kamil Růžička, Ikram Blilou, David Rouquié, Eva Benková, Ben Scheres, and Jiří Friml. Polar PIN localization directs auxin flow in plants. *Science*, 312(5775):883, 2006.

# Using Institutional Frameworks to Conceptualize Agent-based Models of Socio-technical Systems

Amineh Ghorbani, Andreas Ligtvoet, Igor Nikolic, and
Gerard Dijkema

Delft University of Technology,
Faculty of Technology, Policy and Management,
Energy and Industry Group
Delft, The Netherlands
a.ghorbani@tudelft.nl

**Abstract.** Agent-based modeling is a tool that is frequently used to analyze socio-technical systems. The high number of interactions between the agents in these models causes complexity. Different methods are required to understand this complexity and interpret the model behavior and outcomes. In this paper, we apply institutional frameworks, that are also used to analyze socio-technical systems, to understand complex agent-based models. Applying the frameworks to a case study – the Kauffman model – shows that this approach can be successful in giving structure to agent-based models and their outcomes, which can in turn help to interpret the evolving complexity.

## 1 Introduction

Agent-based modeling (ABM) is a popular modeling tool used in different disciplines [1, 7]. One domain where ABM is applied is on the boundary between technical and social sciences. Scientists in this domain are applying ABM to analyze so-called 'socio-technical systems', such as industries and infrastructures. Socio-technical systems are generally defined as complex adaptive entities that require social and technical elements engaged in an environment to reach a goal [11]. Technical artifacts include computers, or machines, whereas social components include actors, organizations, institutions, laws and policies [5, 11].

One of the key characteristics that distinguishes ABM from other types of modeling is the focus on decision-making individuals (agents) rather than the whole system. Since the relationships are not explicitly defined or determined when making the model, the communication

between these individuals results in different networks and multiple outcomes [14]. These networks of interaction between the agents, and the
agents and the environment lead to emergent system structures which
are usually too complex to be interpreted easily.

It is exactly for this interpretation of model outcomes that we turn to
analysis tools that are used by social scientists, institutional economists
and policy analysts. These scientists claim that the complexity in social
systems is especially caused by the positions, relations and behavior of
the parties that own and operate the system [10]. By structuring and
formalizing the analysis of a system in a framework it might be easier
to understand the complex patterns generated. Conversely, the same
frameworks may aid in designing models in the same structured fashion.
In [11], for example, a method to guide the *process* of designing an ABM
is suggested (the so-called 'system decomposition method'). The use of
institutional frameworks may actually formalize designing the *content* of
these models.

Two of the well known institutional frameworks are the four-layer
Williamson model [15] and the institutional analysis and development
framework (IAD) [12]. These frameworks have been successful in explaining behavior and interpreting global outcomes within many different contexts such as economy, organization and policy analysis [10, 6]. The two
main proponents jointly received the Nobel prize for economics in 2009.
Although these frameworks have similar aims, they differ in nature but
can both be very insightful according to the problem domain.

In order to conceptualize the complexity of real world systems and
enable their description as agent-based models, we introduce the two
institutional frameworks. We believe that the frameworks are not only
useful in the design and implementation of agent-based models, but are
also helpful in their analysis. In this paper, we structure an agent-based
model using these frameworks to show how they can be applied. We discuss in which phases of ABM development the institutional frameworks
can be applied and the merits of using the frameworks.

## 2   Institutional Frameworks

The term institution has become widespread in the social sciences in
recent years which reflects the growth in institutional economics and the
use of the institution concept in several other disciplines, including philosophy, sociology, politics [2]. [12] define an institution as 'the set of
rules actually used by a set of individuals to organize repetitive activities that produce outcomes affecting those individuals and potentially
affecting others'. Agreements or rules can be called institutions only if

they are accepted by those involved, used in practice, and have a certain degree of durability [10].

Rules which are formed as a by-product during interactions, are at the heart of institutions. Therefore, institutions can also be considered as set of rules which influence, guide and limit the behavior of actors. They can even be the conscious design behavior by one actor [9]. In other words, institutions have two sides: they enable interactions, provide stability, certainty, and form the basis for trust. On the other side, they codify incumbent power relations and may hamper reform. If institutions fail to fulfill stability or bring about non-decision making and mobilization of bias, there is ground for institutional (re)design [9].

Institutional redesign refers to deliberate changes in institutional characteristics. It is aimed at both the activity of trying to change the institutional features, as well as the content of the institutional change that is aimed for. In order to design institutions, one should be able to understand and analyze institutions and the institutional frameworks that are developed for this purpose. The two well-known frameworks are the institutional analysis and development framework (IAD) by Ostrom and the four layer framework of Williamson which will be discussed in the next two sections.

## 2.1 The four layer Williamson model

The four layer framework of [15] is an approach to describe social and institutional arrangements in an integrated fashion. Like in complex systems theory (see e.g. [4] on the concept of 'panarchy'), each level operates at its own pace, protected from above by slower, larger levels but invigorated from below by faster, smaller cycles. Thus a multi-layer system can be described that shows both bottom-up and top-down causation. Williamson's model is shown in figure 1.

The top level is the 'social embeddedness' level. This is where cultural components such as norms, customs, mores, traditions, and religion are located. These grand institutions change very slowly, in the order of hundreds of years: through continuous interaction with behaviors at the lower levels, this layer is shaped and molded, while at the same time functioning as a brake or anchor for the faster moving levels. One could interpret this level also as the accumulation of the activities at lower levels, thus as emergent from the system. [3], for example, demonstrate that countries may differ fundamentally in values of uncertainty avoidance, individualism, or the relation to authority. At a certain point, certain norms are so embedded in a culture that specific rules are not necessary anymore.

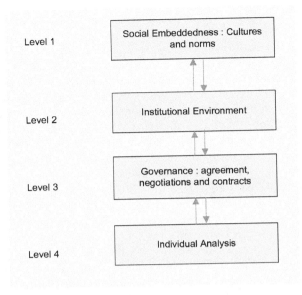

**Fig. 1.** The four-layer model of Williamson

The second level is the 'institutional environment', in which the structures observed are the product of politics and provide the *rules of the game* within which economic activity is organized. Political, legal and governmental arrangements are located here. Laws are an important element of this level. For institutional economists, for example, the laws regarding property rights are an important feature. Often the rules that appear in this level are codified and the outcome of lengthy negotiations. Therefore, change (e.g. rewriting laws) takes place in the order of decades.

On the third level (the 'governance' layer), those arrangements are described that govern the interaction between individuals. Alternative modes of organization are described that range from hierarchical relationships (top-down) to market relationships (completely equal) and a range of complex networked relationships (e.g. joint ventures or co-ops) in between. Typical instruments that economists study in this layer are contractual arrangements, although other interpersonal agreements that are e.g. based on trust could also considered. The institutional environment provides the possibilities and the limitations of the agreements that can be made between the actors (e.g. too much market power is a reason to forbid agreements between large organisations). The period over which decisions are made is of the order of a year to a decade.

The fourth level ('operation and management') moves from the structural to the individual analysis. This is the level with which neo-classical economics is concerned. On the assumption of rationality, individuals calculate their utility and make decisions on variables such as price, demand and output. Individuals focus on getting the marginal conditions right. Adjustments in price and output are made in a more or less continuous way in response to changing (market) conditions.

Transaction cost economics and institutional economics mainly focus on the third or governance level. Here the questions are whether the rules of the game (level 2) and individual behavior (level 4) lead to the required outcomes in terms of market behavior. This framework has also been used by [10] to link the development of complex technical systems to the institutional arrangements, thus combining the social and the technical. What can be seen is that the evolution of socio-technical systems is intertwined: institutional arrangements restrict and steer technical developments, whereas technical innovations require new rules and open up paths to different organizational arrangements.

## 2.2   The IAD Framework

The Institutional Analysis and Development (IAD) framework developed by Elinor [12] is related to Williamson's layers. However, while Williamson's framework allows for more liberty in the analysis of the separate layers, IAD more clearly specifies different elements of the system description (see figure 2).

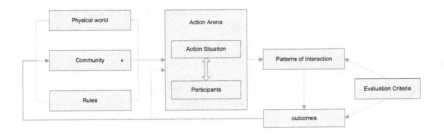

**Fig. 2.** The IAD framework

The central concept is the 'action arena', in which individuals (or organizations) interact, exchange goods and services, engage in appropriation and provision activities, solve problems, or fight. The action arena is described by the participants (who have a set of resources, preferences, information, and selection criteria for action) and the action situation: the actual activity (or 'game') that is to be understood. More detailed description of the action arena is given in section three.

What happens in the action arena leads to patterns of interaction and outcomes that can be judged on the basis of evaluative criteria. The action arena itself is influenced by attributes of the physical world (e.g. climate, present technological artifacts), the attributes of the community in which the actors/actions are embedded (e.g. cultural norms regarding cooperation, demographics or education levels), and the set of rules that the individuals involved use to guide and govern their behavior.

Although physical world and community influence the action arena, it is the rules of the game that actually define it. Therefore, in IAD quite some attention is given to these rules which overlaps with layer two in the Williamson model. Rules are statements about what actions are required, prohibited, or permitted and the sanctions authorized if the rules are not followed. Seven distinct types of rules are distinguished:

**Boundary** Specify who is eligible to play a role: who is in and who is out of the game?

**Position** Determine to what extent a distinction is made regarding the position of the different participants. For example, a buyer or seller on a market have a different role than an auctioneer (and thus different access to information, and different choices).

**Choice** Specify what a participant must, must not, or may do at a specific point of the process.

**Payoff** Assign external rewards or sanctions to particular actions that have been taken.

**Information** Describe what information may or may not be shared by participants and whether they have a set of common, shared information.

**Scope** Define what outcome variables should or should not be affected by the actions undertaken.

**Aggregation** Specify who has responsibility for an action: for example, whether a single participant or multiple participants should come to a decision.

These rules can be analyzed within three distinct layers: the operational, the collective choice and the constitutional choice levels. Like in Williamson's framework, the different levels relate to different time-frames: day-to-day activities fall within the operation level, the collective choices determine what operational activities take place and these are reviewed over a 5-10 year timeframe, whereas the constitutional level determines how the process of collective choice is organized (which is a long-term process).

Apart from Williamson's Level 1 (which in IAD is an exogenous variable, an attribute of community), these three levels quite neatly match the remaining levels in Williamson's framework. For analysis purposes, the emergent patterns are addressed separately in IAD.

Both frameworks also start with the assumption of rational, utility-maximizing individuals but the frameworks are general enough to cover any type of individual behavior. In IAD for example, when activities of actors cannot be explained by the strong assumption of full rationality and complete information, the weaker concept of bounded rationality is introduced [12].

## 3    Case study: Kauffman's Economic Web

In this paper we cover the conceptualization phase of the Kauffman model as a case and describe how we can apply the institutional frameworks to present the concepts of the model.

Applying the two institutional frameworks to agent-based models can be done in three different phases. In the design (conceptualization) phase, the frameworks can be applied to conceptualize the model which in turn would help cover institutional and social concepts in the agent-based model and also guide us in the future analysis of model outcomes. The frameworks can also be used when implementing the actual software model. Even more, we can use the frameworks to analyze the models that have already been run. Of course, ABM is an iterative process and model analysis can lead to modifications in model design and implementation, but what we are suggesting is that the institutional frameworks can actually be applied to any level of these iterations.

### 3.1    Model Specification

The Kauffman model [8] is a simple representation of an economic network that can also be extended to illustrate a complex socio-technical system. This model was originally formulated to show how future wealth would evolve in an economy. We have chosen this model for its simplicity to show how the two institutional frameworks can be applied to agent-based models of socio-technical systems.

The basic model consists of a set of binary strings which represent resources and a grammar table that can convert these strings to other forms . For example if we have '10011' and a conversion rule in the grammar table is: '10 to 111' then the string would be changed to '111101' using the conversion rule in the grammar table.

In [8], Kauffman explains how these simple concepts can represent an evolving economy. The link between the basic terms used in the model and that of the real world are presented in (table 1). From this point, we will use the equivalent names in (table 1) when discussing the model.

To make this economic model more clear, we give a real world example. Suppose that the primary resources are trees in a forest. Factories or individuals use the trees to create new products such as timber, sheets of wood etc with the help of different technologies (e.g. saw). These products can again be used to create newer products (e.g. furniture) by other agents. In this example we can see that the primary resources (trees) can still be available or newly created until the end of the simulation. As the economy evolves, new technologies enter the web which can change the products in additional ways and create an even more complex network.

## 3.2    Extension of the Kauffman model

Since Kauffman does not go into the details of agent interactions, we cannot claim that his is a model of socio-technical systems. The extended version that we conceptualize in this work is aimed to represent a socio-technical system.

In our extension of the Kauffman model, there are agents with certain technology who can use the resources to produce new products; the agents can be considered to be individuals or firms, resources and products are represented as strings. At the beginning of the simulation, these resources are primary - i.e. not produced by anyones else. Later in the simulation they can also be the product of other agents, as it is the main task of agents to convert resources to products. Every tick of the simulation, each agent takes a random turn to look for resources available in the total resource portfolio with desirable characteristics or string patterns and which he has technology for to convert. In the simulation, this equates to manipulate a selection of strings stochastically to produce output. As in a real economy, the agents must survive and remain financially fit. In the model, the agents gain money if other agents use their products and they lose money if their products are not used after some amount of time. New technologies are added to the simulation as time proceeds to resemble how technology may evolve in a real socio-technical system.

Thus, the extended version of Kauffman model represents a socio-technical system according to the following characteristics [13]:

1. Socio-technical systems contain technical components: the grammar table is the representation of technology.
2. Socio-technical systems have changing environment: new rules are added to the model over time to show the evolving technology in a socio-technical system. New primary resources are also added to the system(an example of this in the real world can be trees).
3. Entities who interact with each other and the technical systems: agents use products of other agents and use the technologies to transform strings (make new products).
4. Socio-technical systems are governed by organizational policies and rules: for example, the agents receive points for creating goods that are bought, but lose points if they create unused goods.
5. Socio-technical systems have emergent properties: lifetime distribution and income distribution of agents are in the shape of a power law. The distribution of 1s and 0s in the products also has emergent patterns which show how products evolve over time.
6. Socio-technical systems are non-deterministic: to mimic non-determinism in this model, we used random creation of resources, random turns for the agents to act and random production of the grammar table.

**Table 1.** Representation of a socio-technical system in Kauffman model

| Kauffman feature | Representation of |
|---|---|
| string(1's and 0's) | resources or products |
| conversion rule | technology |
| agent | firm |
| grammar table | set of technologies |
| point | money |

The conceptualization of Kauffman model's extended version using the institutional frameworks is covered in the upcoming section. We will address the other phases of model development in future research.

## 3.3   Understanding the Kauffman model with Williamson's framework

The Williamson framework primarily focuses on structuring and formalizing the drivers of individual (organizations) behavior. With the help of this framework we hope to understand short-term and long-term dynamics of socio-technical systems, and to see where the system is heading subject to higher level setting. Also, how changes in lower levels would affect higher levels of organization and behavior. We use the Williamson framework to structure the different levels of behavior in Kauffman's model by starting with individual behavior which is in the fourth layer.

**Layer 4: Individuals and Interactions** This layer focuses on individual agent behavior, their perception of the world and what they are doing in order to survive. For the Kauffman model, each agent stays alive by producing as many products as he can. He cannot predict what products might actually turn out to be useful; he produces the product without any particular reason and only because he has the technology to do so. Agents take random turns to choose the products of other agents or use primary resources. In this model, there is no specific reasoning for choosing a certain product made by any other agent. In more complicated agent-based models where there are reasons for the decision making process, this layer of the Williamson model would cover all the characteristics of the agents and their individual behavior and decision making strategy.

**Layer 3: Governance Structure** Each agent explores his options with the information he has and decides to interact with other agents in order to gain some benefit. The nature of this interaction which depends on mutual agreement, lies within layer three. This layer covers the agreements and contracts that take place between the agents. It is a form of governance that structures the system and it is based on individual cases of interaction and not something that exists in general within the system. In the Kauffman model, the seller puts his product in a common pool. The choice the seller makes, is random in this model. In many cases of agent based models, there are agreements between agents for certain actions to happen and these types of agreements lie within this layer.

**Layer 2: Institutional Environment** Rules are a fundamental part of ABM. The formal rules of the game lie in this layer. Some examples of rules in the Kauffman model are listed below:

1. If the product of an agent is used by some other agents, the agent gains some certain amount of money.
2. If the product of an agent is not used by other agents after some amount of time (useless product), the agent loses money.
3. If the money of the agent falls below a certain threshold, the agent dies.

In societal systems, these rules are themselves affected by levels below and above, and thus inherently changeable. In an ABM, they are often a given.

**Layer 1: Informal Institutions** This layer deals with the outcome of iterated behavior of individuals and companies. This layer has two implications for agent-based models. First, we can define cultures and norms for the agents (e.g. risk attitudes). Second, we assume that all the global patterns, behaviors and structures that emerge from an agent-based model lie within this layer. The two main characteristics of this layer are that the institutions in this layer take a long term to form, and also, they are very difficult to change. This is similar to how we define emergent properties in an agent-based model. Examples of emergent patterns in the Kauffman case according to our early implementations are:

1. The dying pattern of agents is a power law distribution.
2. The wealth in the economy is a power law distribution.

As illustrated in figure 1, the different layers of Williamson have bidirectional connection. This means that the lower levels influence the higher levels while the higher levels also cause limitations on the lower levels. For example, the characteristics of the individual agents determines their interaction and their actions, in turn, are constrained by the rules of the model, such as the point deduction rule. On the other hand, the cultural level which we can call the emergent institutions level in the case of agent-based models is influenced or formed by all the lower levels of institutions.

This is where the power of the Williamson model lies for the agent-based modeling paradigm. As Williamson suggests, we can arrange all the organizational activities into these four layers and explore how certain norms and cultures have formed. This of course holds for evolutionary models that aim to emulate long term developments and emergence.

As an example of analysis, for the Kauffman model we can change the different levels of institutions separately to see how the two emergent patterns we introduced above have been formed and how they can be changed. For example, changing the local information of the agents can change the wealth distribution since the agents would know whose product not to use in order to decrease that agent's wealth. Or, at the third level we can change the rules of the system to see how they affect the wealth distribution among agents. This perspective of the system can make complex behavior more manageable if not explainable.

## 3.4    Understanding the Kauffman model with the IAD framework

As mentioned previously the IAD framework is aimed at any situations that involve people interacting with each other in a certain context and following certain rules. Each application of the framework focuses on a specific activity (the action situation), the people who take part in this activity (the participants) and the patterns of interactions between them. The combination of activity and participants is called the action arena. The interaction between the stakeholders in an action arena leads to certain outcomes which in turn affect the action arena as well as the exogenous factors influencing the action arena (rules, physical world and community).

The action arena is the most important part of the framework where all the decision making, analysis and prediction, takes place. The action arena chosen for a model depends on the model outcome we are trying to achieve. The chosen arena in the Kauffman model is where agents make decisions to buy goods. Other models may have more than one action arena. According to [12], the different specifications of the action arena are as follows:

**Action Situation** the place where agents interact, is the market where agents buy other agents' products.

1. participants: a number of agents with similar architecture.
2. positions: buyer, seller
3. actions: taking resources from the common pool (buying) putting newly produced products back in the pool.
4. potential outcomes:
   - distribution of wealth among the agents
   - number of dead agents
   - number of products made at the end of the game
   - the max, min, average length of the products produced
5. function that maps actions into realized outcomes:
   - If any agent uses the product of any other agent in the common pool he receives money for producing this useful product.
   - If the products of an agent are not useful(not bought after some amount of time) he loses money.
   - If the agent can not produce goods after some amount of time, he loses money and dies eventually.
6. information
   - The agents have information about the available resources.
   - The agents have information about the technologies they have.
7. cost and benefits assigned to actions and outcomes are the money for production.

**Actors** are defined by the following characteristics:

1. individual preferences
   - Agents choose resources based on whether they have the technology to modify it. The resource could belong to anyone even themselves.
   - The agent may let other agents buy his product or not (randomly).
2. individual information processing capability
   - Agents don't know who the resource they have chosen belongs to, so they have incomplete information.
   - Agents don't know who is buying their product.
3. individual selection criteria
   - How agents choose the technology to make a product from the resource.
   - How agents choose to let other agents buy their product[1].
4. individual resources
   - the technologies
   - The resources available to each agent at a given time. Since we have a common pool, the resources each agent can use is the same as everyone else.

There are several factors affecting the action arena:

**Physical world** is the set of external entities and factors that influence the action arena. For the Kauffman model this includes resources (which are primarily external and later made by other agents) and their properties(e.g. rate of new resources being added to the system, length, pattern etc...).

**Community** is defined by the following characteristics:

1. The norms of the system: Agents can choose any kind of resource they want as long as they have the technology to produce a product from it. Agents don't know the producer of the resource they are using. A product should be put back into the pool as soon as it has been created.
2. The level of common understanding about the action arena: each agent only has information about the available resources at a given time.

---

[1] the selection criteria is random in this example

3. Individual preferences: The preferences among the agents are completely homogeneous in this game, but their set of technologies and thus required resources is different.
4. Distribution of resources: Since we have a common pool the distribution of resources among the members is equal, but the technology the agent has is randomly distributed and some agents may have better technologies than others.

**Rules** [2] that define the play of the game in the IAD framework are:

1. position rules : each agent is a producer
2. boundary rules: each agent stays in the position of a producer until he dies or the game finishes; all agents have the same role, none are excluded
3. authority rules: agents decide on their own product usage and have no hierarchical relationship.
4. aggregation rules: every agents is solely responsible for producing goods. Their success in producing 'useful' goods determines their chance of survival.
5. scope rules: the set of outcomes that may be affected. The only scope rule that affects agent production is that if his money reaches below a certain amount, he dies.
6. information rules:
   - at each node, the agent knows what the available resources are.
   - and which technologies he has.
7. payoff rules: how benefits and costs are required, permitted or forbidden based on the actions taken and outcomes reached.
   - if the resource the agent has chosen is a product of some other agent, the producer receives money for producing a useful product.
   - if it is time to check the age of products in the basket, then check and take away money from those agents who have produced useless products.

These were some of the conceptualizations defined according to the different components of the IAD framework. The components of the framework are optional and are only defined for a specific model if required [12].

---

[2] The rules mentioned here are independent of the rules in the Kauffman model and belong to the specifications of the IAD framework

# 4 Discussion

For the conceptualization of the Kauffman model in this paper, the Williamson and the IAD framework are both insightful when describing socio-technical system with agent-based models. However, in choosing the Kauffman model for its simplicity, we have inadvertently also limited the potential use of the framework for handling this particular model of socio-technical systems. We believe that these frameworks can also be applied to the process of developing agent-based models to give a cultural perspective (behavior emerging from social interactions) to the models and link this perspective to other components. In this section we discuss the advantages and disadvantages of these two frameworks when used in ABM (explained in table 2).

**Table 2.** Comparison of the frameworks

| ABM Phase | Williamson | IAD |
| --- | --- | --- |
| Design | Broad definition of layers, rules and behaviors | Explicitly defines the physical world, distinct types of rules and behaviors |
| Implementation | Provides high level structure for programming | Objects can be defined according to the components of the framework |
| Analysis | Informal layer and outcomes situated in the same level | Specific focus on patterns of interaction and outcomes |

In the *design and conceptualization* phase of agent-based modeling, as we saw in this paper, the two frameworks are both insightful. In Williamson's framework, the researcher can conceptualize the system by thinking in layers and also the interactions the layers may have with each other. For Kauffman's case, first we can focus on the layer of individuals and the properties of each agent such as the initial money or the technologies he may posses. In another layer, we can write down the rules of the game and how for example the pointing system would work. At the global level we can consider the results that we already expect from the system[3] or any form of culture that we are trying to implement (e.g. trust).

While the abstractness of the Williamson framework does not explicitly define the components of the system, it makes the framework suffi-

---

[3] The global patterns of the socio-technical systems one is trying to model may be known in advance as one of the purposes of simulation is explaining the unexplainable behaviors.

ciently generic to be applied to almost any type of model. In the IAD framework on the other hand, the variety of components and details, explicitly defines the key features of the system that may have been missed when designing a system. For example in the Kauffman case, the IAD framework makes us think about the properties of the actors (the *information* each actor has and whether we want different *positions* (buyer, seller) or one *position* as an agent suffices). One other advantage of the IAD framework is the explicit representation of the physical world (e.g. resources in the case of Kauffman model and how long and complex they become as the economy evolves) which has a critical role in agent-based modeling and is not considered in the Williamson framework.

In the *implementation* phase, the Williamson model gives high level of structure for programming (in abstract classes and interfaces) but the IAD framework seems to be more applicable. The components of the framework can be used to define objects since there is a detailed specification of actors. The focus on rule types is also another advantage which can actually help implement the methods of the model.

The *analysis* of an agent-based model is also facilitated with the help of the frameworks, as they both cover the outcome of systems. In the Williamson model, we can change one layer of behavior to see how it would affect the outcomes and have a more structured way of analyzing. For example, for the Kauffman case we can experiment how changing the nature of interactions between agents may actually affect the emergent patterns. Since there is a specific layer (layer one) for the global behavior that emerges from a system, it is insightful to use this model to analyze what level of behavior is causing the emergent patterns. It is worth mentioning that in the Williamson framework, the cultural behaviors are combined with the outcomes in layer one. In the IAD framework there are separate components for patterns of interactions and outcomes beside the community component where the cultures can be defined. On the other hand, different action arenas of a system require separate specifications for the framework which might seem difficult for sophisticated models but pays off in the end for giving a well-structured and concrete model of a system.

In general, the Williamson layers derive their strength from their lack-of definition. They force the researcher to think conceptually in layers, without prescribing exactly what these layers contain. The IAD framework on the other hand, presents the key features of a system that are necessary to be considered in the process of building agent-based models.

Both frameworks are intended for the analysis of (complex) institutional arrangements in societies. As such, they are a useful tool for triggering modelers' thinking about the complex artificial societies they want to create – in the conceptualization phase and, as we hope to show in future work, also in the implementation and analysis phase.

# 5  Conclusion

This paper applies institutional frameworks to conceptualize complex socio-technical systems as agent-based models. This is a new approach which to our knowledge has not been applied to ABM before. This conceptualization gives a more structured approach for framing complex behaviors in agent-based models. The two institutional frameworks can be applied to the design, and potentially implementation and analysis phase of agent-based models. Furthermore, applying institutional frameworks to ABM gives us the possibility of adding a cultural perspective to agent-based models.

Our work is at an early stage. We aim to implement the Kauffman model using the conceptualization we presented in the paper. Consequently, the model will be analyzed according to the frameworks to find out in more detail how they support us in understanding emergent behavior and how we can link them to lower level components.

# References

[1] T.P. Evans and H. Kelley. Multi-scale analysis of a household level agent-based model of landcover change. *Journal of Environmental Management*, 72(1-2):57–72, 2004.

[2] G.M. Hodgson and J. Calatrava. What are institutions. *Journal of Economic Issues*, 40(1):1, 2006.

[3] G. Hofstede and G.J. Hofstede. *Cultures and organizations: Software of the mind*. McGraw-Hill London, 2005.

[4] C.S. Holling. Understanding the complexity of economic, ecological, and social systems. *Ecosystems*, 4:390–405, 2001.

[5] T.P. Hughes, W.E. Bijker, T.P. Hughes, and T. Pinch. The evolution of large technological systems. *The political economy of science, technology, and innovation*, pages 51–82, 1987.

[6] M.T. Imperial. Institutional analysis and ecosystem-based management: the institutional analysis and development framework. *Environmental Management*, 24(4):449–465, 1999.

[7] S. Katare and V. Venkatasubramanian. An agent-based learning framework for modeling microbial growth. *Engineering Applications of Artificial Intelligence*, 14(6):715–726, 2001.

[8] S.A. Kauffman. *Reinventing the sacred: a new view of science, reason and religion*. Basic Books, 2008.

[9]  E.H. Klijn and J.F.M. Koppenjan. Institutional design. *Public manage-ment review*, 8(1):141–160, 2006.

[10] J. Koppenjan and J. Groenewegen. Institutional design for complex tech-nological systems. *International Journal of Technology, Policy and Man-agement*, 5(3):240–257, 2005.

[11] I. Nikolic. *Co-Evolutionary Method For Modelling Large Scale Socio-Technical Systems Evolution*. PhD thesis, 2009.

[12] E. Ostrom, R. Gardner, and J. Walker. *Rules, games, and common-pool resources*. Univ of Michigan Pr, 1994.

[13] I. Sommerville. *Software Engineering*. Addison-Wesley, 8 edition, 2007.

[14] K.H. Van Dam. *Capturing socio-technical systems with agent-based mod-elling*. PhD thesis, 2009.

[15] O.E. Williamson. Transaction cost economics: how it works; where it is headed. *The Economist*, 146(1):23–58, 1998.

# Arguing Validation of Simulations in Science

Fiona A. C. Polack *

Department of Computer Science, University of York, UK YO10 5DD
fiona@cs.york.ac.uk

**Abstract.** Computer simulations are often used in the scientific study of complex systems. However, the validity of simulations is often neglected. This is a particular problem in the potentially-valuable area of agent-based simulation. For collaborative complex systems simulation, the paper draws on conventional simulation validation for inspiration. The proposal is to use validity arguments across multiple validation approaches to express evidence of fitness for purpose.

## 1 Introduction

Scientific study of natural complex systems uses a range of modelling techniques, but the appropriateness of modelling is often left unstated. There is little attempt to express the rationale and reliability of models. It is not that rationale does not exist, but that it is not made accessible – to those engaged in the study of the complex system as well as those who read about the results.

Validation is considered here mainly in relation to agent-based simulations (ABSs), but the issues are of potentially wider relevance. ABSs simulate individual components, with the potential to study processes in complex systems that are not accessible to most other modelling techniques (or laboratory analysis). However, ABSs of natural complex systems have been widely stated to be inappropriate for scientific use, because of the problems of mapping between the scientific context and the simulation (summarised in Polack et al. [27, 26]). The problem is characterised as a lack of perceived rigour leading to a lack of trust. However,

---

* The work presented here is part of the EPSRC-funded CoSMoS project (grants EP/E053505/1 and EP/E049419/1, www.cosmos-research.org/), and refers to the case study from the EPSRC-funded TUNA project (grant EP/C516966/1). The paper draws on work by investigators, affiliates and researchers on the CoSMoS project, notably Paul Andrews, Teodor Ghetiu, Mark Read and Matthew Harbage.

ABSs share conceptual properties with many other forms of modelling: Polack et al. [26] show how other modelling techniques that are no more rigorous or understood than ABSs are trusted by scientists.

This paper outlines some recent work on simulating complex systems, including several examples and the putative CoSMoS process for developing and using complex systems simulations [3]. Conventional approaches to simulation validation are considered in relation to the problem of ABS validation. The paper then proposes the use of *validation arguments*. A validation argument is constructed in relation to the purpose, criticality and impact of the model; it records rationale, and exposes it to external scrutiny.

## 1.1 Previous and related work

The consideration of validation is motivated by existing collaborative exercises involving laboratory scientists and computer scientists. The CoSMoS project has produced ABSs in collaboration with immunologists, plant biologists and others. The basis and philosophy of the CoSMoS project are the subject of a series of papers and a technical report:

- Andrews et al. [3] is a technical report summarising a high-level process for developing complex systems simulations. It focuses on ABS for scientific use, but is potentially applicable to other forms of complex systems modelling. The process is based on observation over a range of collaborative scientific ABS developments.
- Polack et al. [26] reflects on what makes ABS an effective tool in scientific research; it focuses on the roles of scientist and simulation developer, and the way that *trust* is developed by collaborative modelling. The paper also reviews five CoSMoS-related simulation studies.
- Polack et al. [25] identifies an architectural basis for ABS of complex systems, focusing on *components, environment* and *interactions*. The examples include game-of-life cellular automata and hypothetical blood platelets.
- Polack et al. [27] and [28] explore problems with existing ABSs and existing methods for ABS development, and propose aspects of traditional software and simulation engineering that can be used to improve the development, efficacy and validation of ABS. [27] uses the CoSMoS example of lymphocyte migration [2] to illustrate modelling and development concepts.
- Several papers have used argumentation approaches in the context of complex systems modelling. Polack et al. [23] and Alexander et al. [1] look at different aspects of the safety of the hypothetical study of

blood platelets. Ghetiu et al. [10] presents arguments that two ABSs of a plant community are adequately similar, whilst [11] make a case for validation arguments.

There are other robust collaborative simulations used in scientific research. Three examples are outlined here. All are rigorously-engineered simulation examples, which exploit close integration of models in different areas.

*Reactive animation (RA)* [8, 31] is the work of a group led by Harel and Cohen, and has been used to provides detailed models of the behaviour of aspects of the mammalian immune system. The approach combines off-the-shelf tools into a sophisticated and flexible simulation environment. The key diagrammatic modelling components are Rhapsody statecharts (state machines) and Live Sequence Charts (connectivity diagrams). The approach is described as reverse-engineering biological systems into protocols and object-evolution models [7]. The simulation is similar to an ABS: a collection of objects manifest the behaviours defined in the diagrammatic models. Experimentally-derived biological data is used to drive simulations. The simulation can be manipulated directly, through adjustable biological-scale time, zoom, and tracking facilities. It is also possible to adjust the underlying models and see the effects directly on the simulation, which allows the collaborators to experiment with different understandings of the immune biology, and different parameter settings.

A number of groups use stochastic process algebra to construct complementary models of biological networks (see Calder and Hillston [6]). Calder et al. [4, 5] use *PEPA (Performance Evaluation Process Algebra)* to model and analyse biochemical signalling pathways, through a reagent view (akin to state machines) and a network view (a connectivity model). The reagent view can express concentrations and triggers to biochemical product formation, whilst the network view captures time-ordered sequences of events across the system. Whilst diagrammatic views are supported, the PEPA modelling is mathematical, and the two views have been proved isomorphic. The models are not ABSs, but the mathematical approach is closer to an individual-based model than a typical population-level mathematical model. The formalism supports proof of engineering properties such as deadlock-freedom, which improve confidence in the computer models. There is traceability from biological models, at the level of chemical reagents and reaction details. The mathematical models support an algorithmic approach to generating conventional ordinary differential equations, which provides a direct validation link to the scientific analysis of these systems.

Whilst explicit validation activities are not fully reported, like the CoSMoS examples (see [26]) RA and PEPA were developed in interdisciplinary teams, with researchers from several communities bringing complementary skills. The researchers develop mutual understanding of the simulation media and the biological background [26]. Interdisciplinary working also provides hypotheses that the simulation can be used to explore, and, in some cases, confirmatory laboratory experiments for the interpretation of the simulation results.

The third example of systematic modelling and simulation is the swarm robotics work of the Bristol Robotics Lab (www.brl.ac.uk). Unlike the modelling of natural complex systems, swarm robotics benefits from well-researched simulation platforms. Swarm foraging research [18] uses the Player/Stage simulation platform [9], a generic robotics simulator that can be customised to particular robot configurations; its use in swarm simulations is as an ABS in which the robots are agents. A novel aspect of the swarm foraging work is that the foraging behaviour can be formally analysed using a probabilistic finite state machine model [17]. The approach is used to analyse the effect of parameters on the performance of the swarm and to optimise robot parameters. Despite a range of simplifying assumptions (notably homogeneous rather than adaptive robot behaviours), the data analysis shows "excellent agreement between the model and the simulation" [17]. The agreement of analytical and simulation results over a range of settings is taken as validation of the analytic models.

In both PEPA and swarm foraging robotics studies, there is a strong validation link between simulations and analytical models. However, in both cases, there is an additional validation need, since the models (PEPA reagent and network models; swarm Player/Stage simulation) cannot fully represent the environment. This aspect of the validation of the simulation models is currently addressed by observation and expert appraisal.

In collaborative simulation, the design, engineering validation and calibration each represents a process of negotiation and agreement among the participants. The projects reviewed here are notable for their careful engineering integration of methods from computer science, theoretical and laboratory science. The validity of the work is not in question. However, it would be useful to have systematic approaches to record validation that make good practice evident. The problem of validating a model of a complex system arises whatever is modelled and however it is modelled, and not just in the context of ABS for scientific research. This paper focuses on ABSs in scientific research, simply for convenience of presentation.

## 2   The validation problem

In relation to conventional systems simulation, Sargent [33] states that a *model should be developed for a specific purpose... and its validity determined with respect to that purpose.* In complex systems simulation, the necessary level of assurance, or confidence in the validation, also depends on the purpose of the simulation, and, following good software engineering practice, should be set independently of the development of the simulation. A notable feature of complex systems validation is that there can be no absolute notion of validity.

In the CoSMoS project, ABS case studies in a number of scientific contexts have led to some insights into the development and validation of ABSs. An ideal conceptual approach to complex systems validation can be broken down into three separable activities [3]:

- *Engineering validation* appeals to engineering practice. It addresses the quality of construction of a computer artifact. The approach can be characterised as showing that the code meets its specification, and that the specification meets the requirements. Furthermore, engineering validity assumes that code has been adequately verified (i.e. that the program is sound).
- *Calibration* is a tuning activity. The goal is to adjust the ABS parameters, behaviours, scale of operation, etc, to align the simulation with the scientific context. The fine-tuning exposes assumptions, abstractions and simplifications (in relation to the science and the development of the simulation), as well as validating performance and outputs.
- *Scientific validation* comes after engineering validation, in that it assumes that the engineering strengths and limitations of the simulation have been identified, and the simulation has been calibrated. Scientific validation considers the validity of the domain model, for instance by simulating existing real experiments and comparing results.

The three activities, whilst separable, are not necessarily distinct. For example, calibration is not part of a conventional engineering validation process, but may be seen as an engineering activity, since it contributes to the validity of the engineering. In other situations, calibration is part of the scientific validation, because the science is not well-enough understood for the simulator to be calibrated only through engineering-style exploration of parameters.

The overall validity of a complex system simulations can be explored as an argument over the engineering and scientific evidence. The argument is fundamentally tied to the purpose of each simulation, and to the intended criticality and impact of results [3], and is ultimately judged by those using the simulation and its results. *Criticality* relates to the role of the ABS in context. If the simulation is a speculative exploration of possible factors, it should have low criticality. However, if the goal of simulation is to identify a missing link in the scientific understanding, then it has high criticality. *Impact* is similar, but not identical: a non-critical ABS might have disproportionate impact in a new or under-researched area, whilst a high-criticality ABS may have low impact because, once it has identified a critical link, this is confirmed by scientific research. If impact or criticality is high, then explicit engineering and scientific validation must be planned and undertaken. Modelling and design rationale, test results, calibration evidence etc. can be recorded to support the contention that the ABS is an adequate tool for the given purpose. However, where criticality and impact is low, the validation evidence can be more implicit, and could rely on basic engineering quality control and normal laboratory lab-book records.

Simulation design and validation has long been part of conventional systems analysis and engineering. For example, Nance and Sargent [21] review work over four decades, whilst Sargent [33] summarises validation approaches from conventional simulation. The work has been picked up and extended in social science, where the problem of validating an imitation of dynamic behaviour has been widely discussed (e.g. [15, 20]). The following section considers how some existing approaches might relate to complex systems simulation.

# 3  Validation in conventional simulation engineering

Conventional simulation engineering presents a variety of conceptional and practical approaches to validation. For instance, Zeigler [35] presents a theory for modelling and validation of simulations predicated on a homomorphism between conceptual models and simulations. However, he does not show how the homomorphism is established, and the work does not map easily to the complex systems domain. The techniques summarised by Sargent [33] are more practical, drawn from forty years of research on simulation theory and practice [21]. Sargent focuses on validation and verification of a computational model (an implemented simulation) of a problem entity (the subject of simulation, reality). The techniques are summarised in Table 1. The relevance to complex systems simulation validation is considered by allocating Sargent's proposals to three broad categories: comparison-based approaches, testing-related approaches and combined approaches.

**Table 1.** Summary of techniques for validation or verification of simulations from [33]

| Technique | Summary of Sargent's description |
|---|---|
| Animation | Graphical display of operational behaviour |
| Comparison to other models | (1) to analytical model results |
| | (2) to other validated simulations |
| Degenerate tests | Select internal and input parameter values to test degenerate behaviour of simulation |
| Event validity | Event occurrence compared to reality |
| Extreme condition tests | Examine plausibility of model structure and outputs for extreme and unlikely combinations of factors |
| Face validity | Domain experts review design of the system, e.g. internal logic or input-output relations |
| Historical data | Part of a data set is used to build the model, and another part to test it: may be real or purpose-generated data |
| Historical (i.e. traditional) methods | Three of the traditional approaches are: |
| | Rationalism: assumes everyone knows whether underlying assumptions are true; logical deduction from valid assumptions leads to a valid model |
| | Empiricism: every assumption and outcome is empirically validated |
| | Positive economics: requires only that the model can predict the future, so causal relationships, mechanisms, assumptions and underlying structures are of no concern |
| Multistage | Combination the three traditional approaches above: |
| | (i) Develop model assumptions from theory, observation, general knowledge |
| | (ii) Validate assumptions empirically where possible |
| | (iii) Compare input-output relationships of the real and simulated systems |
| Internal validity | Investigation of statistical similarity of results of repeated runs of a stochastic model |
| Operational graphics | Live graphics of performance measures to allow visual assurance of correctness of dynamic behaviours |
| Parameter variability – sensitivity analysis | Change input and internal parameters and validate against similar change in the real system. Sensitive parameters that cause significant change in outputs should be made "sufficiently accurate" before use of the simulation |

| Predictive validation | Compare simulation predictions to reality over time |
|---|---|
| Traces | Trace behaviour of specific entities in the model to determine correctness of model logic and accuracy of results |
| Turing test | Ask knowledgeable individuals to distinguish the real and simulated system outputs |

## 3.1 Comparison-based approaches

Many of the techniques summarised by Sargent [33] rely on comparison with other models or with reality. Sargent considers *Animation* only in the sense of graphical visualisation of operational behaviour, the internals of the simulation. He labels as *Comparison* the validation of a simulation against other valid models, noting both comparison to analytical models and comparison to other validated simulations. *Historical data validation* uses existing or purpose-generated data as a basis for comparison between reality and a simulation, whilst *Predictive validation* compares the output of a simulator to what subsequently occurs in reality. *Operational graphics* is described as visualisation of output such that it can be used to compare behaviour to that measured in other systems. *Turing tests* are the ultimate comparison to reality in which an expert attempts to distinguish the simulation and the real system.

Validation through comparison may rely on visualisation or on data analysis. Of the projects summarised in section 1.1, comparison with analytical models is undertaken in the PEPA simulations, which generate ordinary differential equations for comparison with research-derived equations. The swarm robotics example takes the opposite approach: the analytical model of behaviour is itself validated against the simulations. More generally, ABS complex systems simulations can be compared to other forms of model such as scientific descriptions or mathematical models of population or individual behaviours.

Direct analysis of data in complex systems simulation is inhibited by the stochastic nature of many simulations. Validity needs to be judged by statistical measures, and gives only a level of confidence that the simulation results and real data are comparable.

Visualisation, whilst not essential, is widely used in simulation of complex systems. Validation is commonly undertaken by asserting the visual similarity of the simulation and the subject of simulation. In one sense, visualisation of a complex system simulation does provide validation, since simulations normally seek to show that some required emergent behaviour occurs when the system behaviour at the low level is observed at a higher level. However, visualisation alone is inadequate. Some of the reasons for this are:

- Visually similar behaviours arise from different underlying processes.
- A simulation abstracts from reality, such that a single feature in the simulation represents a range of features in the real system; visual behaviours may actually be artifacts of the abstraction.
- Unless great care is taken in abstraction and implementation, the behaviours that should be emergent may be coded in to the simulation.

In relation to comparison to existing models, there are few scientific simulations that have adequately-documented validity. Ghetiu et al. [10] explore ways to show that two simulations are adequately similar. An important caveat to validation by comparison (and, indeed to validation in relation to complex systems in general) is that the comparison can only show the extent to which the targets are similar. The validation cannot tell us anything about the independent validity of either model.

Another significant caveat to complex systems comparison is that the model never includes all the complexity of reality. A strict approach to modelling complex systems might expect to start at the very bottom, such that classical physics emerges from quantum mechanics, chemistry from classical physics, and so on. Indeed, Lloyd [19] comments that a complete simulation of a natural complex system is a quantum computer that efficiently simulates the Universe. However, a rational view is that, at each level of interest, the effects of lower levels can be aggregated or omitted without a significant effect on the desired emergent behaviour. What this means for validation is that any parameter or behaviour that is included in the model is a surrogate for a vast number of parameters and behaviours that are not in the simulation.

## 3.2   Testing-related approaches

Many of the validation techniques summarised by Sargent [33] are derived from testing approaches used in systems and software engineering. Sargent provides simple examples to explain most of these approaches. For example, in a job-queue structure, a typical *Degenerate test* would look at the behaviour of the queue when the rate of arrival of jobs is greater than the service rate. In relation to *Extreme condition tests*, Sargent uses the example of an inventory system: "if in-process inventories are zero, production output should usually be zero" [33]. In relation to *Parameter tests and sensitivity analysis*, Sargent refers to quantitative

and qualitative testing of internal and input parameters – this is similar
to practices such as range or domain testing in software engineering.

Face and internal validity and traces validation are also derived from
testing. *Face validity* is stated to be an expert analysis of the validity
of the model and its behaviour in terms of review of features such as
simulation logic or input-output relationships. *Trace validity* focuses on
analysing the behaviour of entities in the model (e.g. agents in an ABS).
*Internal validity* uses statistical techniques to explore the consistency of
results from repeated runs of a stochastic simulation [33]. *Event validity*
combines a testing approach with a comparison approach: the example
in [33] is a simulation of a fire department, in which the number of fires
in the model is compared to the real situation.

In complex systems simulation, approaches derived from conventional
software testing can be used both in the engineering validation and in
calibration of the simulation. Validation needs to explore the behaviour
of the simulator under a wide range of inputs and operating conditions,
in a similar way to that in which testing challenges software. A typical
problem in testing a complex system simulation is that the range of
possible interactions and the requirement for emergent behaviour makes
it impossible to accurately predict system behaviours. For all but the
most trivial throw-away simulation, calibration is an essential part of the
analysis of the simulation behaviour. In relation to calibration, sensitivity
analysis has been proposed (see [32]), but little agreement on appropriate
techniques[1].

Testing and related challenges to a complex system simulation have
to be applied to all aspects of the simulation. In an ABS, engineering validation and calibration should be applied to the agents (low-level
components) and to the environment (without agents); then, realistic
collections of agents need to be tested in the environment, in a suitably
representative range of possible situations. There is a known danger here,
that the calibration activities tune the parameters and behaviours so that
the simulation gives visually appropriate results. This merely determines
a set of parameter values and behaviours that produce a nice picture;
it does not validate the simulation. A visually-accurate result may arise
from behaviours that are unrelated to the subject of simulation.

In the complex system simulation examples reviewed in section 1.1,
the PEPA and swarm robotics cases include mathematical analyses which
clearly depend on a correct interpretation of parts of the system logic
and relations (face validity). However, as noted above, formal analysis
of a complex system often requires significant simplifying assumptions.

---

[1] In relation to CoSMoS, Read et al. are addressing sensitivity analysis for
ABS; initial ideas are noted in [30, 29].

Furthermore, face validity, like Turing test comparison, relies on evaluation by a domain expert, but simulations of complex systems in scientific contexts are generally undertaken to explore the possible processes and behaviours of systems that are not adequately understood: the necessary expertise required for evaluation does not exist. The exploratory nature of complex systems simulation also limits use of techniques for validating the internal structure, data and behaviour of the simulation, such as event validity. The lower-level components (agents) may be conventional engineered systems [24, 25] that can be validated conventionally, but the components must also be validated against reality. Abstraction typically replaces a multi-layer complex reality with two or three layers. Validation can never be definitive because the correspondence between abstract concepts and reality is indirect.

Sargent's internal validity [33], which uses statistical techniques to explore the consistency of results from repeated simulation runs of stochastic simulations, is highly relevant to validation of complex systems simulation, since the simulations are stochastic and results from one run cannot be considered to be representative. Statistical analysis is an essential part of calibration and sensitivity analysis. The techniques that are used to explore consistency across runs of one simulation are also applicable when comparing simulation outputs to the real system. However, as with other attempts at validation of observed behaviour, statistically-similar results do not say anything about the validity of the processes and input data that give rise to the results.

In relation to complex systems analysis, care has to be taken with statistics. Data (from models or reality) are rarely amenable to parametric analysis, and it is easy to generate spurious results through inappropriate use of statistics. In critical situations, it might be necessary to seek expert advice on statistical analysis – whether in relation to internal validity, comparison, empirical validation or any other data-based validation approach.

## 3.3   Combined approaches

Traditional validation approaches are individually insufficient for validation of a complex system simulation. For conventional simulations, multistage validation [22] combines the historical methods identified by Sargent [33]: the approach establishes the theoretical credibility of the simulation, identifying and validating the assumptions, and validating assumptions and theory-use through empirical research, as well as assessing whether the output of the simulation is as expected.

The multistage approach raises issues in complex system simulations:

- There is little relevant theory against which to assess theoretical credibility.
- In relation to assumptions, it is potentially possible to record all identified assumptions made in representing the domain and in developing the simulation. It is common for calibration to uncover unexpected simulation behaviour, which may be due to hidden assumptions. It is impossible to know whether a complete or sufficient set of assumptions has been captured.
- Empirical validation of assumptions (etc) suffers from the surrogacy problem: the simulation is not usually a close match to reality.

### 3.4   Traditional approaches: summary

In summary, no single technique is sufficient on its own to determine that a complex systems simulation is valid, but most of the approaches considered have a potential contribution. Combined approaches are promising, and could incorporate more techniques than the multistage recommendation above [33, 22]. Combination goes some way to addressing the problems of applying any individual validation approach to complex systems simulation, and is clearly required to address the different validation activities: engineering validation, calibration and scientific validation; validation of components, environment and components in the environment, and so forth.

There is a need for circumspection in ascribing objectivity to the outcomes of any validation activity, because its contribution can only be considered in the context of all the activities carried out. In complex systems simulation, validation needs to present an argument across a range of techniques, applied to more than just the parameters and outputs of the simulation. We need to consider carefully what it might mean for a complex systems simulation to be valid.

## 4   An example highlighting some validation challenges

In traditional computing, if a system operating within its specification meets its requirements then it is said to be valid. Validity is contingent on operating context, but within context validity can be asserted in absolute terms.

When considering complex systems (in general and as the subjects of simulation), specification of behavioural requirements is problematic. A typical complex system simulation is an abstract model of components, but the real components may exist at many levels of abstraction, and the system that is being simulated may be poorly understood.

In a traditional computer development, the interaction of components can be precisely specified, but in a complex system, emergent high-level behaviours are the result of enabling, rather than dictating, appropriate interaction, not only among components but also between components and the environment. Complex behaviours are not simple combinations of low-level behaviours, but include behaviours that are apparent only when the system is observed at a higher level than that of the components.

In engineering terms, this means that validation of complex system simulations must consider at least (in order of increasing difficulty):

- the components;
- potential component interaction;
- the environment, from the perspective of its potential interaction with components;
- the means of identifying the occurrence of appropriate emergent behaviours.

In addition to engineering validation, the simulations must also be validated with respect to the reality that is being simulated.

The range of validation activities needed can be illustrated by considering the CoSMoS lymphocyte migration simulation [2]. Briefly, lymphocytes are white blood cells with a key role in immune processes. They enter a lymph node from the blood circulation by a series of state changes ("capture", "rolling", "migration"), through chemical interaction with the vesicle walls. The simulation activity started with simulation developers and scientists working together to identify appropriate scientific abstractions for the simulation. A model could have been constructed at the biochemical (or even physics) level, but the questions of interest to the scientists relate to how changes in vesicle shape might influence the rate of migration, so it was appropriate to abstract to generic lymphocyte state changes. Validation of this abstraction involved discussion and agreement with the scientists. Essentially, validation required the scientist to "sign off" the abstract model – although in reality there is ongoing dialogue and iterative improvement of the model.

In the lymphocyte migration study, the abstract scientific model (the "domain model", see [3]) maps cleanly to design and implementation models for components (agents) [2, 27], so the engineering validation of components was straightforward. The design of the environment, though straightforward in engineering terms, is almost impossible to validate in any scientifically-meaningful way. Some of the problems encountered are as follows.

- Immunologists are not experts in blood circulation or vesicle architecture, so the scientific guidance is limited (the study was not of sufficient importance to involve a wider group of scientific experts). This means that the simulation environment is a gross simplification of the dynamics of the circulatory system, and the effects of the simplification are not amenable to validation within the scope of the study.
- Empirical data is gathered from live subjects; data across subjects varies by several orders of magnitude for some parameters. Scientists are used to working with approximate data, but driving a simulation with such uncertain data is interesting.
- The dynamic behaviour of the immune system cannot be studied directly in a single subject, so the data to drive the ABS, and empirical results that could be compared to simulation results, come from a large number of different subjects. Scientists understand the potential and limitation of interpolation across point data from different subjects, but there is no research on how multi-subject data and behaviour can be compared with a simulation that implicitly represents the immune system of one subject.

The lymphocyte migration study shares with all ABS the problem of surrogacy: that the included features are also surrogates for excluded features. Surrogacy became even more problematic when the simulation was elaborated from data-only output to a visualisation. Empirical science provides approximate rates for state transitions in the migration process, and these are appropriate for driving the data-only simulation. However, in the visual form, each lymphocyte has a spatial location. Lymphocyte migration is related to adjacency to the vesicle wall which is now an explicit feature of the environment. There is a question as to whether it is acceptable to continue using the empirical data to drive the simulation. The approach taken was simply to record that the inconsistency exists, both in development and in analysis of results. Collaborative review of the simulation results suggests that the anomaly does not have a significant effect, but this cannot be validated objectively.

## 5 Argumentation and critical systems engineering

The discussion of conventional and complex systems validation highlights the need to combine validation techniques, and the residual uncertainty that is inevitable in validation of complex systems simulations. We suggest that the optimal way to address validation is to construct *validity arguments*.

The technique of argumentation is used in critical systems engineering to present a case to regulators/certifiers for believing that a system has a required properties, most commonly safety [1, 14]. It is impossible to absolutely demonstrate properties such as safety; instead evidence is collected based on criteria such as use of accepted development practices; software, system and sub-system testing; mechanical analysis; past experience; cumulative usage outcomes; and field trials. The evidence is used to support an argument that the risk associated with a system is *As Low As Reasonably Practicable* (ALARP), within the operational environment for which the system is designed. Kelly [14] describes the general approach to constructing and documenting safety cases; this forms the basis for much commercial and military safety assurance, and assurance of other critical systems qualities.

The origins of argumentation in critical systems is not unlike the situation in complex systems simulation. Early safety-critical systems were unregulated and potentially unsafe [16]. Consequent deaths from accidents led to regulation, part of which is usually certification. Potentially-dangerous systems are allowed if there is sufficient evidence that they are safe to operate in the given context. Evidence used to be based on process ("I have followed good engineering practice, so my system is safe"). This approach is not only unsatisfactory, but also inhibits innovation by limiting engineers to use of approved processes.

A significant improvement in safety management came with product-based certification. Independent regulators set the safety criteria and specific evidence requirements that systems must meet. Developers establish evidence and tie it together by means of a structured argument, known as a safety case. It is still possible to cite an approved process as evidence, but this evidence is relegated to an appropriately-subordinate role. A safety case is accepted or rejected based on independent review of its arguments and evidence. Acceptability is not an absolute, and can change over time, in the light of experience or new evidence.

Critical systems arguments present parallels to scientific investigation, particularly in biology, where understanding of complex natural systems is a developing area, with much debate and competing theories. If the validity of complex simulations can be established through argument, then it can be reviewed when new scientific understanding arises, or when a simulation is improved. Revisiting the validation argument reveals the extent to which validity still holds. The extent and formality of the validation exercise depends on the purpose of the simulation and the nature of the collaboration underpinning the simulation.

## 5.1    Some possible validation arguments for ABS

The goal of ABS validation is to demonstrate *reasonable adequacy*. Just
as it is possible to argue that a system is as safe as reasonably practicable,
ABS validation must be an appropriate justification of the trust placed
in the simulation and its results. Validation is thus an argument that a
simulation meets its scientific and engineering objectives. As in safety engineering, the argument is open to scrutiny by whoever needs convincing.
If an ABS supports laboratory research into a biological complex system,
then reasonable ABS adequacy is comparable to the adequacy of results
of laboratory experimentation [26].

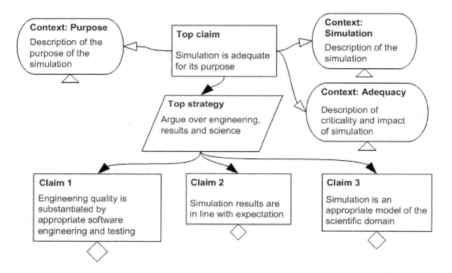

**Fig. 1.** The top two levels of a generic argument that a simulation is adequate
for its intended purpose: for GSN notation, see text

Figure 1 shows the start of a generic argument of adequacy, using the
goal structuring notation (GSN) [14]. The top goal or claim represents
the intention of the argument: to show that the simulation is adequate for
its purpose. Here, the generic top-level claim is addressed using a strategy
(Top strategy) that leads to definition of three sub-claims. The diamond
decorations indicate that these claims need further consideration. GSN
provides notations for recording the context, assumptions and justifications for a claim. Figure 1 has three contexts recorded, namely the
descriptions of the simulation, the purpose and the criticality and im-

pact of the simulation. The triangle decorations show that these contexts need instantiating, linking to the relevant statements or documents.

Ghetiu et al. [11] propose an alternative generic argument that maps more directly to the conventional characteristics of an engineering validation: addressing the top-level claim that *Simulation results are valid according to the research purpose and requirements*. The proposed subgoals are: (i) that the conceptual model is valid; (ii) that the simulator accurately implements the model; and (iii) that the experimental setup is adequate (i.e. that it supports the needs of calibration and the scientific purpose).

The argument itself needs validating. One way to do this is to record justifications for the strategy. In figure 2, the justification simply identifies the individuals (who would be named) who agree that this strategy is appropriate. It would be possible to provide more than one strategy, perhaps addressing different people's concerns in relation to the adequacy of the simulation.

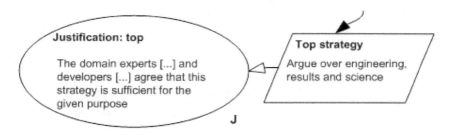

**Fig. 2.** Recording the justification of a strategy using GSN notation

The extent to which each claim in an argument is elaborated is dependent on the purpose, criticality and impact of the simulation, and on the people who review the argument. These aspects cannot be separated: some people are inherently harder to convince than others, but some simulations are more critical than others, so there is a trade-off of conviction against importance.

To illustrate validation in more detail, consider the argument extract in figure 3, recording that the results of a simulation are adequately similar to those obtained in the laboratory. This is adapted from an argument, presented in [10], that seeks to validate a simulation by showing that it is adequately similar to an earlier simulation that features in ecological literature (the remainder of the argument in [10] shows the extent to which the two simulations capture the same scientific model and the same implementation model).

The validation argument in figure 3 lays out what has been (or should be) done to show that the results of the simulation adequately replicate the results of existing scientific experiments. This is a typical comparison activity, in which the simulation is calibrated against known scientific results. The argument includes two new pieces of notation. The lowest-level claims represent a set of experiments, each of which must be satisfactorily evaluated: the set is indicated by the black dot on the link from the strategy to Claim: experiment n – the dot can be annotated with the number of experiments to be conducted, and the number would need to be related back to the set of canonical experiments identified under context: results. The circular component at the bottom of the diagram represents a solution, the evidence that supports or substantiates a claim – here the evidence is that, for the first experiment, the simulation results are acceptably similar to the results from laboratory experiments.

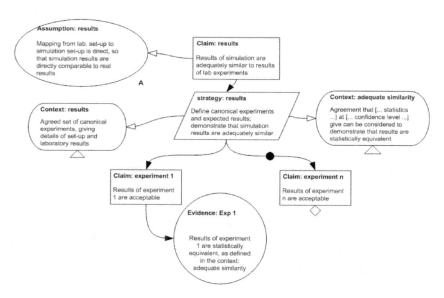

**Fig. 3.** Arguing that results are adequately equivalent: for GSN notation, see text

The argument of result adequacy is spelt out quite carefully, with reference to particular statistics and confidence levels. Context: results points to the need to identify and agree a set of canonical experiments.

However, there are assumptions which may or may not be acceptable to the reviewers of the argument; three of these are as follows.

- The explicit assumption, Assumption: results states the need for a sufficiently close mapping between the real and the simulated experiments: it does not say anything about how this is demonstrated. The reviewers of such an argument might want to see an explicit mapping argument for each experiment.
- Context: adequate similarity would give details of the equivalence metric. It might be necessary to state the number of runs required for each experiment, or to provide other detail of statistical testing. It might also be necessary to provide a justification for the particular statistic used.
- The outcome of experiment 1 is stated in Evidence: Exp 1; it might be necessary to provide a more explicit set of evidence and, perhaps, a link to the statistical analysis.

Conversely, it is possible that a reviewer would accept the argument that results are adequately similar based on no more than is presented here – without even instantiating the detail of the two contexts. A reviewer who was convinced by arguments of quality in the engineering and scientific areas (the parts of the argument not shown here) might be inclined to take results evidence as relatively unimportant, merely seeking to know that the experiments had been properly thought out. The unresolved Claim: experiment n implicitly records the assumption that all other results were adequate, and make it easy to identify where to strengthen the argument if necessary.

Whilst in most cases an informal agreement that an argument is sufficient is acceptable, a more structured analysis of the argument may be needed. The evaluation of goal-structured arguments is not a well-researched area. However, systematic deviational analysis can be used to seek weaknesses (see [13, 34] for examples of deviational analysis applied to models).

# 6 Discussion

The novel contribution to validation is in the presentation of the argument of validity. Argument over evidence is an accepted way to demonstrate the qualities of critical systems such as safety or dependency. Furthermore, the justification steps that contribute to a validity argument are closely allied to the internal rationalisation that takes place in conducting science (or any other activity where resource usage must be justified!). However, the use of explicit argumentation techniques in validation is novel. The principle is similar to critical systems usage: where a quality cannot be demonstrated absolutely, the basis of believing that the quality pertains to the system must be demonstrated as fully as possible, with reference to evidence.

This paper focuses on validation of ABS in scientific contexts. However, it has potentially wider application – to other forms of computational model, and to other scientific instruments. Consider the three forms of enhancement that Humphreys [12] describes, in relation to scientific instrument use:

**Extrapolation:** the extension of an existing modality. For example, vision is extended via a microscope.
**Conversion:** the conversion of a feature from one mode to another. For example, a visual display on a sonar device.
**Augmentation:** the extension of accessibility to features which, in their natural form, cannot be detected by humans. Examples include the detection of intangibles such as magnetism and particle spin.

The enhancements form a continuum, and are not discrete: conversion is often used with extrapolation and augmentation. The conceptual division is useful in focusing attention on how and why people use some instruments and not others. ABS in the scientific study of complex systems is primarily an augmentation tool – it allows a scientist to explore behaviours that are not detectable or not explicable under laboratory conditions. However, there are many more augmentation tools in this field, and similar validation issues apply in all cases.

Candidate techniques for validation of complex system simulations range from simple visual comparisons to elaborate statistical analyses. The choice of techniques for a particular simulation must relate to the purpose of the simulation, and to its apparent criticality and impact. As in other critical systems argumentation, validity arguments are intended to record the basis of belief and to expose the basis of belief to external scrutiny. In the case of validation of complex systems simulation, it may not be necessary to present a complete argument or a complete set of evidence – the rigour and completeness of the argument should match the intended purpose, and the criticality and impact of the simulation. However, it may also be necessary to revisit the argument repeatedly, as the understanding of the science modelled in the simulation changes or improves, or if the impact of the simulation is greater than that originally assumed, or to satisfy those who must be convinced by the argument. It may be appropriate to turn to other critical systems techniques to challenge the simulation models more thoroughly, or it may be sufficient merely to demonstrate informally that a reasonable effort has been made in relation to assumptions and justifications.

In critical systems engineering, argument patterns may be used to express common strategies. Validation arguments or parts of argument can also be generalised to patterns suited to different modelling contexts. Patterns could also address the ways in which different techniques can be adapted and combined in complex systems simulation validation.

As in uses such as safety case argumentation, the validity argument establishes the basis for trust [26]: if a claim cannot be substantiated, or is demonstrably false, the argument (and possibly the simulation) must be revisited. However, whereas in safety case argumentation, an unsubstantiated claim invalidates the top claim of system safety, in a validation argument, claims may be left as only potentially substantiatable. There are many reasons for this: complex systems simulations are exploratory, testing and uncovering new hypotheses; scientific analysis of complex systems is still exploratory, and many details of mechanisms and interactions are not well known; some areas (such as immunology, neurology) do not admit direct access to the living system so dynamics are at best hypothetical; the mappings from continuous reality to a digital computer simulation are not well understood. The validation argument is thus likely to be incomplete, because the scientific and engineering basis of the simulation is incomplete. Validation is as much about the limitations of the simulation as it is a demonstration of adequacy.

# 7   Summary

Simulation is an essential part of scientific study of complex systems. ABS are not widely trusted, despite some successful examples of collaborative simulation development. The part of the problem of uptake that is addressed in this paper is validation: the proposals could be applied to any form of modelling or simulation of complex systems.

Having identified the goals and scope of the validation problem, conventional validation approaches are reviewed, and their appropriateness for complex systems simulation is discussed. The need to validate many aspects of a simulation (components, environment, interaction) and to address both engineering and scientific aspects of validation, points to the use of arguments of validity. This is illustrated using GSN argument fragments.

Work on validity argumentation is ongoing within the CoSMoS project. Research is addressing patterns and guidelines for validity argumentation, particularly in relation to calibration. In addition, arguments are

being used in comparing simulations, and in comparing simulations to
published research results.

# References

[1] R. Alexander, R. Alexander-Bown, and T. Kelly. Engineering safety-critical complex systems. In *Workshop on Complex Systems Modelling and Simulation*, pages 33–63. Luniver Press, 2008.

[2] P. S. Andrews, F. Polack, A. T. Sampson, J. Timmis, L. Scott, and M. Coles. Simulating biology: towards understanding what the simulation shows. In *Workshop on Complex Systems Modelling and Simulation*, pages 93–123. Luniver Press, 2008.

[3] P. S. Andrews, F. A. C. Polack, A. T. Sampson, S. Stepney, and J. Timmis. The CoSMoS Process, version 0.1. Technical Report YCS-2010-450, Dept of Computer Science, Univ. of York, 2010. www.cs.york.ac.uk/ftpdir/reports/2010/YCS/453/YCS-2010-453.pdf.

[4] M. Calder, S. Gilmore, and J. Hillston. Modelling the influence of RKIP on the ERK signalling pathway using the stochastic process algebra PEPA. *Transactions on Computational Systems Biology VII*, 4230:1–23, 2006.

[5] M. Calder, S. Gilmore, J. Hillston, and V. Vyshemirsky. Formal methods for biochemical signalling pathways. In *Formal Methods: State of the Art and New Directions*. Springer, 2008.

[6] M. Calder and J. Hillston. Process algebra modelling styles for biomolecular processes. *Transactions on Computational Systems Biology XI*, 5750:1–25, 2009.

[7] S. Efroni, D. Harel, and I. R. Cohen. Reactive Animation: realistic modeling of complex dynamic systems. *IEEE Computer*, 38(1):38–47, 2005.

[8] S. Efroni, D. Harel, and I. R. Cohen. Emergent dynamics of thymocyte development and lineage determination. *PLoS Computational Biology*, 3(1):0127–0135, 2007.

[9] B. Gerkey, R. T. Vaughan, and A. Howard. The Player/Stage project: Tools for multi-robot and distributed sensor systems. In *International Conference on Advanced Robotics*, pages 317–323, 2003.

[10] T. Ghetiu, R. D. Alexander, P. S. Andrews, F. A. C. Polack, and J. Bown. Equivalence arguments for complex systems simulations - a case-study. In *Workshop on Complex Systems Modelling and Simulation*, pages 101–140. Luniver Press, 2009.

[11] T. Ghetiu, F. A.C. Polack, and J. Bown. Argument-driven validation of computer simulations – a necessity rather than an option. In *VALID*, 2010. to appear.

[12] P. Humphreys. *Extending Ourselves: Computational Science, Empiricism, and Scientific Method*. Oxford University Press, New York, 2004.

[13] J. A. Clark J. Srivatanakul and F. A. C. Polack. Stressing security requirements: Exploiting the flaw hypothesis method with deviational techniques. In *Symposium on Requirements Engineering for Information Security*, 2005.

[14] T. P. Kelly. *Arguing safety – a systematic approach to managing safety cases.* PhD thesis, Department of Computer Science, University of York, 1999. YCST 99/05.

[15] G. Küppers and J. Lenhard. Validation of simulation: Patterns in the social and natural sciences. *Journal of Artificial Societies and Social Simulation*, 8(4), 2005.

[16] N. Leveson. High-pressure steam engines and computer software. *IEEE Computer*, 27(10):65–73, 1994.

[17] W. Liu, A. F. T. Winfield, and J. Sa. Modelling swarm robotic system: A case study in collective foraging. In *Towards Autonomous Robotic Systems*, pages 25 – 32, 2007.

[18] W. Liu, A. F. T. Winfield, L. Sa, J. Chen, and L. Dou. Towards energy optimization: Emergent task allocation in a swarm of foraging robots. *Adaptive Behavior*, 15(3):289–305, 2007.

[19] S. Lloyd. *Programming the Universe: A Quantum Computer Scientist Takes On the Cosmos.* Vintage, 2005.

[20] S. Moss. Alternative approaches to the empirical validation of agent-based models. *Journal of Artificial Societies and Social Simulation*, 11(1), 2008.

[21] R. E. Nance and R. G. Sargent. Perspectives on the evolution of simulation. *Operations Research*, 50(1):161–172, 2002.

[22] T. H. Naylor and J. M. Finger. Verification of computer simulation models. *Management Science*, 14(2):B92–B101, 1967.

[23] F. Polack. Argumentation and the design of emergent systems. Working paper, available at www-users.cs.york.ac.uk/~fiona/PUBS/Arguments.pdf.

[24] F. Polack and S. Stepney. Emergent properties do not refine. *ENTCS*, 137(2):163–181, 2005.

[25] F. Polack, S. Stepney, H. Turner, P. Welch, and F. Barnes. An architecture for modelling emergence in CA-like systems. In *European Conference: Advances in Artificial Life*, volume 3630 of *LNAI*, pages 433–442. Springer, 2005.

[26] F. A. C. Polack, P. S. Andrews, T. Ghetiu, M. Read, S. Stepney, J. Timmis, and A. T. Sampson. Reflections on the simulation of complex systems for science. In *ICECCS*, pages 276–285. IEEE Press, 2010.

[27] F. A. C. Polack, P. S. Andrews, and A. T. Sampson. The engineering of concurrent simulations of complex systems. In *CEC*, pages 217–224. IEEE Press, 2009.

[28] F. A. C. Polack, T. Hoverd, A. T. Sampson, S. Stepney, and J. Timmis. Complex systems models: Engineering simulations. In *ALife XI*. MIT press, 2008.

[29] M. Read, P. S. Andrews, J. Timmis, and V. Kumar. A domain model of Experimental Autoimmune Encephalomyelitis. In *Workshop on Complex Systems Modelling and Simulation*, pages 9–44. Luniver Press, 2009.

[30] M. Read, P. S. Andrews, J. Timmis, and V. Kumar. Using UML to model EAE and its regulatory network. In *International Conference on Artificial Immune Systems*, volume 5666 of *LNCS*. Springer, 2009.

[31] A. Sadot, J. Fisher, D. Barak, Y. Admanit, M. J. Stern, E. J. A. Hubbard, and D. Harel. Towards verified biological models. *IEEE/ACM Transactions on Computational Biology and Bioinformatics*, 2007.

[32] A. Saltelli, K. Chan, and E. M. Scott, editors. *Sensitivity Analysis: Gauging the Worth of Scientific Models*. Wiley, 2000.

[33] R. G. Sargent. Verification and validation of simulation models. In *37th Winter Simulation Conference*, pages 130–143. ACM, 2005.

[34] T. Srivatanakul. *Security Analysis with Deviational Techniques*. PhD thesis, Department of Computer Science, University of York, 2005. YCST-2005-12.

[35] B. P. Zeigler. A theory-based conceptual terminology for M&S VV&A. Technical Report 99S-SIW-064, Arizona Center for Integrative Modeling and Simulation, 1999. www.acims.arizona.edu/PUBLICATIONS/publications.shtml.

# Adapting Gosper's Hashlife Algorithm for Kinematic Environments

William M Stevens

Department of Physics and Astronomy, Open University, Milton Keynes, UK, MK7 6AA

**Abstract.** Gosper's *hashlife* algorithm is adapted and applied to a three-dimensional kinematic environment by simulating the environment using interleaved simulation periods, each of which can be modeled using a different set of cellular automaton rules. Information about the global state of the environment is stored in order to decide when to make a transition from one set of cellular automaton rules to another. The adaptations are described in the context of a specific environment but can be applied to other similar environments.

## 1  Introduction

Bill Gosper's ingenious *hashlife* algorithm [6] was originally applied to Conway's Game of Life. Gosper pointed out that it could be extended to other geometries, dimensions and number of states per cell. The algorithm has recently been applied to other cellular automata (CAs) using the open source Golly CA simulator [1]. Self-replicating programmable constructors (SRPCs) that were too large to implement or simulate using the technology available at the time of their conception have recently been simulated using the *hashlife* algorithm [10, 9, 4, 7].

Usually when a constructing machine embedded in a cellular automaton environment needs to make a new structure it does so by using CA rules that are designed to allow empty space to turn into a component part on demand: there is no "conservation of matter". One reason for this is that the CA environments that these systems are embedded in do not have any model of motion: components and machines are fixed in place.

If we introduce the constraints that component parts cannot be created or destroyed or transformed into different types of component part then we must also provide a mechanism by which component parts can be moved from one location to another.

If a constructing machine cannot create component parts on demand then it must obtain them from its environment. Either the machine must be mobile and able to seek component parts within its environment, or individual parts within the environment must move around so as to encounter the constructing machine which can then make use of them.

Additionally a constructing machine must be capable of positioning component parts correctly relative to other parts in the machine being constructed. Either the machine being constructed must be manoeuverable or the constructing machine must contain a subsystem (perhaps the whole machine) which is capable of moving a component part to a specific location.

These considerations lead to a strong case for bestowing mobility not only upon component parts but also upon larger structures. Once a means of moving structures in space is provided, we are faced with the problem that a structure may fall apart when it is moved unless some means of connecting neighbouring component parts is also provided.

These are among the considerations that led to the environment described in Sect. 2.

## 2    A 3D Kinematic Environment with 6 Part Types

This section describes a discrete space, discrete time 3D kinematic simulation environment called CBlocks3D. This is a development of the previously published 2D CBlocks environment [13], having a greatly reduced set of component parts and not requiring any ability to create component parts out of empty space. It can be regarded as an implementation of the kinematic environment originally proposed by von Neumann [5]. There are 6 different types of part in the CBlocks3D environment: a signal propagation part (the *wire* part), a signal processing part (the *nor* part), a part for moving other parts (the *slide* part), a part for rotating other parts (the *rotate* part), a part for connecting other parts (the *fuse* part) and a part for disconnecting other parts (the *unfuse* part). These part types are chosen so as to be as simple as possible whilst spanning the operations required for construction within the environment. Boolean signals can pass between neighbouring parts. It takes one time unit for a part to respond to an input signal. In a single time unit, a part may move one unit in any one of six directions under the action of a *slide* part. When a part moves, all parts directly or indirectly connected to it also move.

## 2.1   Describing parts

Standard notation from set theory is used in this section. Readers unfamiliar with this notation should consult reference [8] or [2].

We define six direction vectors

$$EAST = (1,0,0), \quad WEST = (-1,0,0),$$
$$NORTH = (0,1,0), \, SOUTH = (0,-1,0),$$
$$FRONT = (0,0,1), \quad BACK = (0,0,-1).$$

$D$ denotes the set of these vectors

$$D = \{EAST, WEST, NORTH, SOUTH, FRONT, BACK\}.$$

We define the function

$$\text{opposite}((x,y,z)) = (-x,-y,-z).$$

$L$ is the set $\{True, False\}$, and $T$ denotes the set of part types

$$T = \{wire, nor, slide, fuse, unfuse, rotate\}.$$

A part is completely described by the tuple

$$(P.location, P.primary, P.secondary, P.type, P.output, P.connect)$$

where

$$P.location \in \mathbb{Z} \times \mathbb{Z} \times \mathbb{Z},$$
$$P.primary \in D,$$
$$P.secondary \in D \setminus \{primary, \text{opposite}(primary)\},$$
$$P.type \in T,$$
$$P.output \in L \times L \times L \times L \times L \times L \text{ and}$$
$$P.connect \in L \times L \times L \times L \times L \times L.$$

$P.location$ is a 3-tuple $(x,y,z)$ that specifies the location of the part.

Two vectors are needed to specify the orientation of a part $P$ in three dimensional discrete space. The primary axis $P.primary$ is a vector that lies on the line from the centre of $P$ to the centre of one face of $P$ (this face is referred to as the *active face* of $P$). The secondary axis $P.secondary$ is perpendicular to the primary axis. For example, in Table 1 the primary axis of each part points up the page ($NORTH$) and the secondary axis points to the right of the page ($EAST$).

The notation $X[Y]$ is used to refer to the $Y$th element of the tuple $X$. It is convenient to use the direction vectors $D$ to index the $P.output$

and *P.connect* tuples, so we define that the vectors $NORTH$, $EAST$, $SOUTH$, $WEST$, $FRONT$ and $BACK$ can be used to index the 1st, 2nd, 3rd, 4th, 5th and 6th elements of a tuple respectively.

$P.connect[d] \in L$ where $d \in D$ specify the connectivity state of $P$. If a part $P$ is connected in a particular direction $d$ to a neighbouring part $Q$ then $P.connect[d] = True$ and also $Q.connect[$opposite$(d)] = True$. If a part $P$ is not connected to its neighbour $Q$ which lies in direction d, then $P.connect[d] = Q.connect[$opposite$(d)] = False$. If a part $P$ has no neighbour in direction $d$ then $P.connect[d] = False$. When a part is moved by a *slide* part all parts connected to that part also move. A moving part $P$ will also push a neighbouring part $Q$ that lies in the direction of motion of $P$, even if $Q$ is not connected to $P$. A part can be rotated by a *rotate* part regardless of its connectivity state. When a part is rotated by a *rotate* part its connectivity state remains unchanged.

Boolean signals can pass between the faces of neighbouring parts. Neighbouring parts do not need to be connected in order for signals to pass between them. Each face of a part acts either as an input or as an output. It takes one time unit for a signal to propagate from a part's inputs to its outputs or for a part to respond to signals at its inputs. When we talk of the value of a signal at an input face of a part, this is the value being output by an abutting face of a neighbouring part, or *False* if there is no abutting face.

$P.output[d] \in L$ where $d \in D$ are the outputs of $P$. So for example if we have an isolated *nor* part $P$ with $P.primary = EAST$, the values of its outputs will be $P.output[EAST] = True$ and $P.output[d] = False$ for all other $d \in D$.

Table 1 describes the function of each type of part. A graphical representation is shown for each part type. In Table 1, the letters $N$, $E$, $S$, $W$, $F$ and $B$ are used to refer to the value of signals at the $NORTH$, $EAST$, $SOUTH$, $WEST$, $FRONT$ and $BACK$ inputs of a part. Any output not specified in Table 1 has the value *False*. In these diagrams $NORTH$ is up the page, $EAST$ is to the right of the page and $FRONT$ is out of the page. Note that the parts are shown here in one particular orientation and the function of each part is described with respect to this orientation. The Boolean ¬ (negation) and ∨ (OR) operators are used in Table 1.

The *wire*, *nor* and *rotate* parts have rotational symmetry about their primary axis and can therefore be in any one of 6 functionally distinct orientations. The *slide*, *fuse* and *unfuse* parts have no rotational symmetry and can therefore be in any one of 24 distinct orientations.

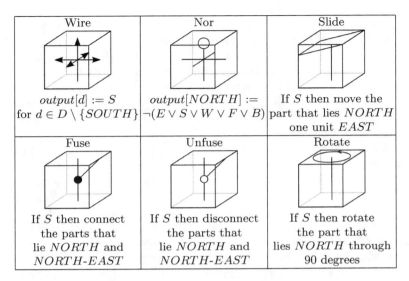

| Wire | Nor | Slide |
|---|---|---|
| $output[d] := S$ for $d \in D \setminus \{SOUTH\}$ | $output[NORTH] :=$ $\neg(E \vee S \vee W \vee F \vee B)$ | If $S$ then move the part that lies $NORTH$ one unit $EAST$ |
| Fuse | Unfuse | Rotate |
| If $S$ then connect the parts that lie $NORTH$ and $NORTH$-$EAST$ | If $S$ then disconnect the parts that lie $NORTH$ and $NORTH$-$EAST$ | If $S$ then rotate the part that lies $NORTH$ through 90 degrees |

**Table 1.** Part types in CBlocks3D.

All parts except the *nor* part have a single input which is at the $SOUTH$ face in Table 1. The *nor* part has 5 inputs at the $EAST$, $WEST$, $FRONT$, $BACK$ and $SOUTH$ faces in Table 1.

A self-replicating programmable constructor made from approximately 60,000 parts was implemented in this environment [12, 15]. The SRPC collects parts from a disorganised collection in the environment and then uses those parts to construct any specified machine, with self-replication as a special case. The need to simulate this SRPC on inexpensive hardware in a matter of days rather than weeks was the primary motivating factor for developing an efficient simulation algorithm for the CBlocks3D environment.

A brief description of an important subsystem of the SRPC can be found in reference [14]. This subsystem is capable of identifying any part presented to it.

## 3 Simulating the CBlocks3D Environment

When the CBlocks3D environment was first devised, a simulator was developed that iterated over every part in the environment at each time step to calculate the effect that each part had on other parts. The simulator was later refined so as to simulate only those parts likely to change state from one time step to the next.

The fact that structures made from connected parts can move together in a single time step means that the environment cannot be directly implemented in a CA. In a CA the state of a cell one time step into the future is affected by cells in a finite neighbourhood. In the CBlocks3D environment a moveable structure can have any length in any dimension. As a consequence of this one end of an arbitrarily long rod can end up moving because the far end of the rod was acted upon by a *slide* part in the previous time step.

Given that the CBlocks3D environment cannot be implemented using a CA and that there is no difficulty in using other methods to simulate it, there would be no cause for further consideration of CA were it not for the existence of the *hashlife* algorithm. The potential performance improvement that the *hashlife* algorithm offers, coupled with the knowledge that several structures in the SRPC being simulated have a high degree of spatial and temporal regularity, which *ought* to be capable of being simulated in a more efficient way, strongly motivated further consideration of whether CA techniques could be adapted to simulate the CBlocks3D environment.

Although the possibility of arbitrarily sized moving structures means that in general the environment cannot be implemented using a CA, it is perfectly possible to simulate periods of time during which no movement occurs using a CA. One possible end point of such a period is shown in Fig. 1. Here a *slide* part is activated, and via its action on part $A$ it will cause structure $B$ to move one unit to the right in the next time step. In the first simulator that was written for the CBlocks3D environment an algorithm was written to work out which parts in a structure move when any part in the structure is acted upon by a *slide* part. This algorithm worked by propagating movement information from one part to another in a similar way to the way that information is propagated in a CA. This led to the realisation that a separate set of CA rules can be used to propagate movement information from one part to another, and another set can be used once all movement information has been propagated to carry out the movement operation.

# 4   An Algorithm Based on Gosper's *Hashlife*

Firstly we devise a set of CA transition rules for carrying out all operations except movement of parts in the CBlocks3D environment. These rules will allow any simulation to be carried forward to a state where movement of parts is about to occur. This set of transition rules is referred to as the NORMAL set.

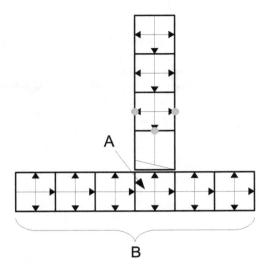

**Fig. 1.** Structure B is about to be moved by the *slide* part.

Then we devise a second set of CA transition rules that allow information about movement to propagate from one part to another. This set is referred to as the PROPAGATION set. A third set of transition rules is used to perform movement. This set is referred to as the MOVEMENT set.

After this we show how information about the global state of the environment can be used to decide when to switch from one set of rules to another. The *hashlife* algorithm is then modified to allow results calculated from all three sets of CA rules, as well as state information for subregions of the environment, to be incorporated into the same data structure.

There are some possible conflicts in these rule sets: it is possible for two parts to be acted upon by both a *fuse* part and an *unfuse* part at the same time, and for two or more active *rotate* or *slide* parts to act upon the same part. Further discussion of these conflict situations and how they may be dealt with is deferred to Sect. 5.

The NORMAL set of CA rules is by far the most complex of the three. It is given as a set of If-Then rules below.

Two parts are *neighbours* when they have abutting faces. Two parts are *diagonal neighbours* when they share a single edge. Graphical examples are shown beneath each rule. The graphical examples are not intended to cover all possibilities encompassed by each rule, but rather to aid the reader in understanding the rules. Small gray circles are used to indicate *True* outputs. In the rules that deal with connectivity a small gap between parts is used to show that they are not connected. In the rule concerning the *slide* part, a grey circle with an arrow in it is used to indicate a part that is marked for movement.

In these rules $P$ refers to a part occupying a cell. These rules show how $P$ changes state from time $t$ to time $t+1$. If no rule is applicable then a cell remains unchanged. Any outputs of $P$ not explicitly mentioned in these rules are set to *False*.

If $P.type = wire$ and the input of $P$ faces an output of a *wire* or *nor* part outputting a *True* signal then $P.output[d] = True$ for $d \in D \setminus \{opposite(P.primary)\}$ at $t+1$. Otherwise these outputs will be *False* at $t+1$.

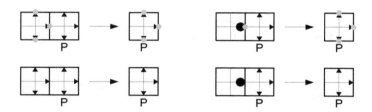

If $P.type = nor$ and any of the inputs of $P$ face an output of a *wire* or *nor* part outputting a *True* signal then $P.output[P.primary] = False$ at $t+1$. Otherwise $P.output[P.primary] = True$ at $t+1$.

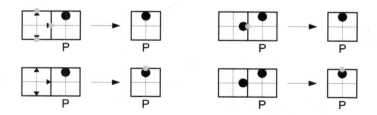

If $P$ has a *rotate* part $R$ as a neighbour and $R.primary$ points toward $P$ and the input of $R$ faces an output of a *wire* or *nor* part outputting a *True* signal then $P$ will rotate through 90 degrees anticlockwise about $R.primary$ by $t+1$. Note that the state of the $P.connect$ tuple is not changed by a rotation.

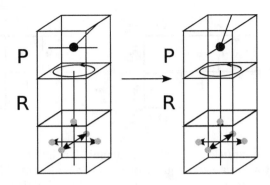

If $P$ has a *fuse* part $F$ as a neighbour and *F.primary* points toward $P$ and the input of $F$ faces an output of a *wire* or *nor* part outputting a *True* signal and $P$ has a neighbour $Q$ in the direction of *F.secondary* then at $t + 1$ $P$ and $Q$ will be connected.

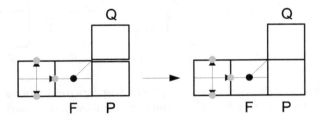

If $P$ has an *unfuse* part $F$ as a neighbour and *F.primary* points toward $P$ and the input of $F$ faces an output of a *wire* or *nor* part outputting a *True* signal and $P$ has a neighbour $Q$ in the direction of *F.secondary* then at $t + 1$ $P$ and $Q$ will be disconnected.

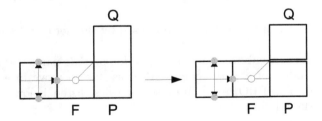

If $P$ has a *fuse* part $F$ as a diagonal neighbour and *F.primary* points toward a neighbour $Q$ of $P$ and the input of $F$ faces an output of a *wire* or *nor* part outputting a *True* signal and *P.location* − *Q.location* = *F.secondary* then at $t + 1$ $P$ and $Q$ will be connected.

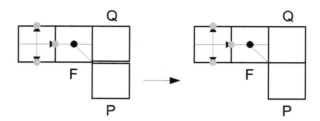

If $P$ has an *unfuse* part $F$ as a diagonal neighbour and *F.primary* points toward a neighbour $Q$ of $P$ and the input of $F$ faces an output of a *wire* or *nor* part outputting a *True* signal and *P.location* − *Q.location* = *F.secondary* then at $t+1$ $P$ and $Q$ will be disconnected.

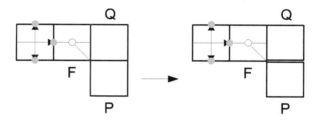

If $P$ has a *slide* part $S$ as a neighbour and *S.primary* points toward $P$ and the input of $S$ faces an output of a *wire* or *nor* part outputting a *True* signal then mark $P$ as a part to be moved in the direction of *S.secondary*.

The shape of the neighbourhood used by this set of rules is shown in Fig. 2.

The last rule above concerns the action of *slide* parts and introduces the possibility of a part being marked as due to move in a particular direction. The PROPAGATION set of rules, listed below, propagates this marking information.

If $P$ is connected to neighbour $Q$ and $Q$ is marked as due to move in direction $d$ then mark $P$ as due to move in direction $d$.

If $P$ has a neighbour $Q$ in direction $e$ and $Q$ is marked as due to move in direction $d = \text{opposite}(e)$ then mark $P$ as due to move in direction $d$.

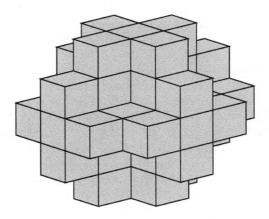

**Fig. 2.** The shape of the neighbourhood used by the NORMAL set of rules.

The MOVEMENT set of rules then uses the marking information to carry out movement:

If a cell $C$ has a neighbour $P$ in direction $e$ and $P$ is marked as due to move in direction $d = \text{opposite}(e)$ then $P$ will be moved into $C$.

Both the PROPAGATION and MOVEMENT sets have a 3D von Neumann neighbourhood.

In order to decide when to switch between one set of CA rules and another we need to know the following information about the environment:

**Condition 1** There is at least one part in the environment that is marked as due to move.

**Condition 2** Running the PROPAGATION set of rules will not result in any changes (i.e. propagation is complete).

Simulation proceeds using the NORMAL set until Condition 1 becomes true. Then simulation proceeds using the PROPAGATION set until Condition 2 becomes true. Then a single iteration of the MOVEMENT set is run, before switching back to the NORMAL set.

A readable explanation of Gosper's algorithm is given in reference [11]. Extrapolating directly from Gosper's 2D algorithm, a 3D *hashlife* algorithm for a CA with a 3D Moore or von Neumann neighbourhood uses an octree data structure to represent the state of a universe, where a level-$n$ node in the tree represents a cube $2^n$ cells on a side. Each node contains pointers to eight sub-cubes, each $2^{n-1}$ cells on a side. Whenever two or more level-$n$ nodes would represent identical configurations of cells, a single node is used with two or more pointers at level-$n + 1$ pointing to the same node. A RESULT pointer either points to nothing, or else to a cube $2^{n-1}$ cells on a side containing the result of simulating the inner cube $2^{n-2}$ steps into the future. Whenever a node is formed, a hash value is calculated based on its contents and it is stored in a hash table for future use should an identical node be formed in future. The main advantage conferred by the hash table is that nodes within it are likely to contain already-computed RESULT pointers.

Whereas Gosper's algorithm simulates $2^{n-2}$ steps forward for a level-$n$ cube, we cannot do this for the CBlocks3D environment. One reason for this is that the NORMAL rule set has a neighbourhood with some cells two units away from the central cell. Therefore calculations of the future state of level-0 nodes may require information from neighbours that are not visible until level-3. For this reason the 4 by 4 by 4 cube that is the result of simulating an 8 by 8 by 8 cube is one time step into the future of the larger cube (rather than 2 as would be the case for a Moore or von Neumann neighbourhood). Furthermore if we are stepping forward in time using the NORMAL rule set at a rate of $k$ time steps per octree calculation step then at some time between $t$ and $t + k$ we may need to switch to the PROPAGATION set. We need to detect when this occurs, back-track to time $t$ and then step forward more slowly until we reach the time when we need to switch to the PROPAGATION set. To reduce the amount of back-tracking that is required, we limit the distance into the future that the algorithm can calculate to 16 time units by ensuring that a doubling of the simulation time-step-size only occurs at levels 4,5,6 and 7 (whereas in *hashlife* doubling occurs at every level). Because of the above mentioned need to back-track and step forward slowly, we also allow results to be calculated with no time-doubling at all, so that simulation can proceed at a rate of one time step per octree calculation step.

Condition 1 and Condition 2 mentioned above can be calculated in a hierarchical way by storing these conditions in each node, and then combining them to calculate the conditions for a higher level node.

These considerations lead to the following structure for a single octree node, expressed using the C programming language:

```
#define EMPTY_FLAG      1
#define NOTPROP_FLAG    2
#define PROPSTATIC_FLAG 4
```

```
#define CALCINDEX_CALC   0
#define CALCINDEX_CALC2  1

typedef struct otNode
{
    unsigned char level;
    unsigned char flags;
    otNode *calc[2];

    union
    {
        block leaf;
        struct otNode *children[2][2][2];
    } u;

    struct otNode *next;
} otNode;
```

The *level* member is not strictly necessary, since so long as the level of the highest level node is known, any algorithm operating on a universe can work out which level it is operating on.

The *flags* member has bits for recording whether or not the node is empty (EMPTY_FLAG), whether or not Condition 1 is true (NOT-PROP_FLAG) and whether or not Condition 2 is true (PROPSTATIC_FLAG). The set/reset sense of each of these flags is chosen so that flags from sub-nodes can be combined by a bitwise AND operation.

The *calc* member is used to store the result of simulating either one application of a CA rule set (CALCINDEX_CALC) or multiple applications (CALCINDEX_CALC2). Note that it is not necessary to have separate members of *calc* for the NORMAL, PROPAGATION and MOVEMENT rule sets because for a given configuration of cells only one of these rule sets will result in a change. Specifically, the NORMAL rule set will never be applied to any configuration of cells containing cells marked for movement. The PROPAGATION and MOVEMENT rule sets, when applied to any configuration of cells not containing any cells marked for movement, will result in no change. The PROPAGATION rule set will also result in no change when applied to a configuration of cells where movement information is already fully propagated. When the application of a rule set to a configuration is known to result in no change the *calc* member is not used, instead the unchanging result is formed from the central sub-sub-nodes of the node in question.

For level-0 nodes the *leaf* member contains all information about the state of a single cell. This information fits into 18 bits as follows:

Part type: 3 bits
Part orientation: 5 bits
Output state: 1 bit
Connectivity state: 6 bits
Movement propagation direction: 3 bits

For higher level nodes the *children* member contains pointers to subnodes.

The *next* member is not part of the representation of an environment, but is used for making lists of nodes which are used both in the hash table and also for keeping track of unused nodes.

# 5   Discussion

## 5.1   Conflicting operations

The CA rule sets described in Sect. 4 do not behave in a consistent way when an attempt is made to carry out two or more conflicting operations on a part at once, such as attempting to move a part in two different directions at the same time. The original implementation of the CBlocks3D environment did cater for these situations, and they were dealt with as follows:

- Any attempt to both connect and disconnect two parts at the same time results in no change in their connectivity state.
- Any attempt to rotate a part about two different axes at the same time results in no change in its orientation.
- Any attempt to move a part in two different directions at the same time results in no movement of the part in question – effectively the movement operations that caused the conflict are cancelled.
- Any attempt to move two or more parts into a single empty cell results in no movement of the parts in question – effectively the movement operations that caused the conflict are cancelled.

The two movement-conflict situations listed above are more subtle than they may at first appear. A comprehensive discussion of the issues involved was carried out by Arbib [3]. Arbib attempted to devise a resolution strategy that is more complex than the one described above and which aims to produce behaviour that more closely approximates the motion of rigid bodies under Newtonian mechanics within the limits of a discrete space environment. For example, Arbib stipulated that an attempt to move a part in two orthogonal directions would succeed and would result in the part moving diagonally. His detailed analysis ends with the sentence 'I hope this formulation is contradiction free', so it is not clear whether he succeeded.

Clearly there is more than one choice of conflict resolution strategy and the extent to which movement-conflicts can be efficiently handled within a given simulation framework depends upon the choice of strategy, which in turn depends upon the reason why a system is being simulated.

One of the simplest strategies is not to resolve movement-conflicts or any other conflicts at all, but instead to declare, as part of the definition of the CBlocks3D universe, that conflicts are illegal and that no well designed mechanism should cause them to arise. This was done in the context in which the CBlocks3D environment was first used [12]. This approach is adequate when using the environment to simulate well designed mechanisms with predictable behaviour, but would not be suitable if one were simulating a system with unpredictable behaviour where conflicts could not be ruled out.

## 5.2  Performance

The original simulator for the CBlocks3D environment, mentioned in Sect. 3, was capable of simulating the SRPC system at an average rate of about 50 iterations per second on the hardware available at the time. At this rate simulation for a complete replication cycle (220 million iterations) would have taken over six weeks. On the same computer system, using 3.5 Gb of RAM for the hash table, a *hashlife*-based algorithm was able to carry out the same simulation in 10 days at an average rate of 250 iterations per second. The algorithm has been improved since this simulation was carried out, so the same simulation would now be expected to take about 4-5 days.

Table 2 shows the effect that varying the memory available to the algorithm has on performance. This table is based on simulating the first 2 million time steps of the operation of the SRPC mentioned at the end of Sect. 2.

The complete source code for the algorithm described in this paper is available from:

http://www.srm.org.uk/downloads/CBlocks3DHash.zip

| Number of nodes available | Memory used | Iterations per second |
|---|---|---|
| $2 \times 10^6$ | 88 Mb | 125 |
| $4 \times 10^6$ | 175 Mb | 201 |
| $6 \times 10^6$ | 263 Mb | 272 |
| $8 \times 10^6$ | 350 Mb | 313 |
| $10 \times 10^6$ | 439 Mb | 345 |
| $15 \times 10^6$ | 658 Mb | 396 |
| $20 \times 10^6$ | 877 Mb | 428 |
| $30 \times 10^6$ | 1316 Mb | 559 |
| $40 \times 10^6$ | 1754 Mb | 563 |
| $50 \times 10^6$ | 2193 Mb | 620 |

Table 2. Effect of varying the number of nodes on simulation speed.

# 6    Acknowledgements

Thanks to the anonymous reviewers of this paper for valuable suggestions. Most of the work in this paper was undertaken as part of an
external PhD research project with the Department of Physics and Astronomy, Open University, UK.

# References

[1] Golly: a cross-platform application for exploring john conway's game of
life and other cellular automata. http://golly.sourceforge.net (visited on
14th june 2010).

[2] Online wikibooks chapter on sets, http://en.wikibooks.org/wiki/set_theor
y/sets, (visited on 14th june 2010).

[3] M. A. Arbib. Machines which compute and construct. In *Theories of
Abstract Automata*, pages 355–361. Prentice-Hall, Englewood Cliffs, New
Jersey, 1969.

[4] W. R. Buckley. Signal crossing solutions in von Neumann self-replicating
cellular automata. In *Automata 2008*, pages 453–501. Luniver Press,
Frome, UK, 2008.

[5] A. W. Burks and J. von Neumann. *Theory of Self-Reproducing Automata*.
University of Illinois Press, Urbana, Illinois, 1966.

[6] R. W. Gosper. Exploiting regularities in large cellular spaces. *Physica D*,
10(1-2):75–80.

[7] T. J. Hutton. Codd's self-replicating computer. *Artificial Life*, 16(2):99–
117, 2010.

[8] D. L. Johnson. *Elements of logic via numbers and sets*. Springer-Verlag,
London, 1998.

[9] R. Nobili. The cellular automata of John von Neumann. http://www.pd.infn.it/~rnobili/wjvn/index.htm (visited on 14th june 2010).

[10] U. Pesavento. An implementation of von Neumann's self-reproducing machine. *Artificial Life*, 2(4):337–354, 1995.

[11] T. G. Rokicki. An algorithm for compressing space and time. *Dr Dobb's Journal*, 31(4):12–18, 2006.

[12] W. M. Stevens. *Self-replication, construction and computation*. PhD thesis, Open University, Milton Keynes, UK.

[13] W. M. Stevens. Simulating self replicating machines. *Journal of Intelligent and Robotic Systems*, 49(2):135–150, 2007.

[14] W. M. Stevens. Parts closure in a kinematic self-replicating programmable constructor. *Artificial Life and Robotics*, 13(2):508–511, 2009.

[15] W. M. Stevens. A self-replicating programmable constructor in a kinematic simulation environment. *Robotica*, (in press).

# A Reflection on Complex Systems: Interesting and Challenging

Paul S. Andrews[1], Teodor Ghetiu[1], Tim Hoverd[1],
Jenny Owen[1], Adam T. Sampson[2], Douglas N. Warren[2], and
Antonio Gomez Zamorano[1]

[1] University of York, York, UK, YO10 5DD
[2] University of Kent, Canterbury, UK, CT2 7NF
psa@cs.york.ac.uk

Research carried out on the CoSMoS project[3] has touched on various aspects of complex systems modelling and simulation. This includes numerous and varied case-studies (see [3] for a summary), methodology of simulation construction [1], and the development of tools and techniques for simulation construction [2]. The content of this abstract provides a summary of the experiences of working with complex systems by the core team who have carried out the research. The aim is to provoke further thought and discussion on a range of issues regarding complex systems. The abstract is structured around two main themes: researching complex systems is ultimately rewarding and *interesting*, but can prove to be *challenging*.

First we highlight a number of general issues that make working with complex systems an interesting and rewarding activity:

**Questioning:** the types of question asked about a complex system in order to construct models and simulations can often challenge domain knowledge. Such questions might examine aspects of the domain that are not typically the subject of focussed research.

**Collaborations:** researching complex systems is a truly transdisciplinary activity. Not only do researchers from different domains collaborate to investigate complexity, but various engineering techniques are employed to construct tools for scientific inquiry. Collaborations also allow the cross-fertilisation of insights into the nature of complexity that arise in different disciplines.

**Engineering:** the development of tools and techniques to investigate complex systems can be directly applicable to more engineering activities. Large-scale engineered systems, such as the Internet, are becoming

[3] Funded by EPSRC grants EP/E053505/1 and EP/E049419/1, and a Microsoft Research Europe PhD studentship. See http://www.cosmos-research.org

increasingly complex themselves, and will benefit from complexity research.

**Tools for Discovery:** simulations of complex systems can be used both to study specifically posed question, and for generating hypotheses that can be used to explore the real system. Often simulation tools can be used in an ad-hoc manner to discover interesting and emergent properties of the system.

**Insights:** we can often gain insights into complexity where similarities are observed between examples of complex systems that are not obviously related. A defining quality of complexity science is the understanding gained of the subject through observation of complex systems examples. We also appear naturally drawn to visual examples of systems with non-trivial structures and dynamics, even if they are simply engineering artefacts.

We now outline below a number of general issues that have been challenging and hindered our work with complex systems:

**Terminology:** "complex systems" and associated terms such as "emergence" and "validity", have, to a large degree, evaded rigorous definition. This can hinder communicating research, hide the real meaning of results, or conceal a lack of understanding of the system under study.

**Uncertainty:** due to their very nature, the complex systems we study often contain many more uncertainties than certainties. This makes modelling a difficult art that involves a large set of assumptions to be made. Claims made about complex systems research results often fail to properly account for assumptions and uncertainties.

**Scalability:** it can often be difficult to approach the level of scale (for example the number of entities) required to study a complex system using a simulation. This can be complicated by complexity arising at differing and multiple scales.

**Communication:** interaction between different roles in developing complex systems simulations is vital, but often difficult to achieve. This is amplified when communicating work across disciplines. There can also be a lack of openness in research (failure to release computer code and document all tools and techniques) that can hinder reproducibility.

**Reflection:** it is rare to see introspection in the world of complex systems that questions what we are doing and the approaches we are taking. Such introspection should address the hard issues highlighted above, providing guidance based on personal experience.

In this abstract we have outlined in brief a series of issues that we have found both difficult and rewarding when working with complex systems.

The purpose is to reflect on the nature of complex systems research from our personal experience, which may serve as a starting point to consider what the future of complex systems research could be.

# References

[1] Paul S. Andrews, Fiona A. C. Polack, Adam T. Sampson, Susan Stepney, and Jon Timmis. The CoSMoS process version 0.1: A process for the modelling and simulation of complex systems. Technical Report YCS-2010-453, Department of Computer Science, University of York, 2010.

[2] Tim Hoverd and Adam T. Sampson. A transactional architecture for simulation. In *ICECCS 2010: Fifteenth IEEE International Conference on Engineering of Complex Computer Systems*, pages 286–290. IEEE Press, 2010.

[3] Fiona A. C. Polack, Paul S. Andrews, Teodor Ghetiu, Mark Read, Susan Stepney, Jon Timmis, and Adam T. Sampson. Reflections on the simulation of complex systems for science. In *ICECCS 2010: Fifteenth IEEE International Conference on Engineering of Complex Computer Systems*, pages 276–285. IEEE Press, 2010.

# Modelling the Role of Chromosome Missegregation in Cancer Therapies

Arturo Araujo[1], Peter Bentley[2], and Buzz Baum[3]

[1] UCL CoMPLEX
[2] UCL Computer Science
[3] UCL MRC Laboratory for Molecular Cell Biology
a.araujo@cs.ucl.ac.uk

In traditional chemotherapy, anti-cancer drugs are used to target and kill rapidly dividing cells. However, only a fraction of the cells die from the treatment. In order to reduce the size of the tumour, repeated doses must be administered [3]. By simulating this procedure in a model that accounts for chromosome missegregation as the main drive for cancer progression new insights can be developed.

Previous work created an individual-based model that exhibits emergent cancer-like behaviour through the interaction of abstracted cancer genes in cells [1]. Throughout the simulation, individual cells may divide, die or remain alive, depending on the interactions between the number of abstracted genes that make up cells and the conditions of the system (Fig 1.a). Each cell has a genome (Fig 1.b and Fig 1.c), composed of 3 kinds of genes distributed across homologue chromosomes: Apoptosis Regulatory (AR), Cell-Division Regulatory (CDR) and Chromosome Segregatioin Regulatory (CSR) Genes, interacting through the algorithm described in Fig1.a. If dividing, the cells have a probability of missegregating entire chromosomes and thus generating daughter cells with abnormal DNA content and gene expressions. Previous work demonstrated that through chromosome missegregation events, the cells evolve cancer-like genotypes with unlimited proliferation and avoidance of death [1]. This is achieved by abstracting oncogene activation (increased number of copies of cell-division regulatory genes) and the loss of tumour suppressor genes (decreased number of apoptosis regulatory genes).

It is of importance, both biologically and clinically, to investigate the effects that chromosome missegregation and aneuploidy – the cellular state of having an abnormal number of chromosomes – have in cancer drug treatments [2]. Making use of this framework, in silico experiments were carried out to simulate the interaction between cancer treatments and chromosome missegregation on two configurations of cancer-like systems (Fig. 1.b and Fig1.c). A cancer drug that kills all cells entering in mitosis for 2 consecutive time steps was simulated. In *Treatment A*, the

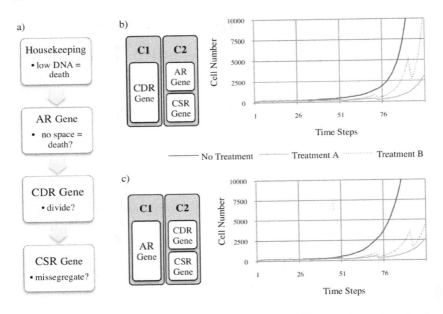

**Fig. 1.** a) Algorithm for each time step b) Gene Configuration I genotype leads to a cancer-like behaviour. Treatment A is administered at time steps 30, 50 and 70. Treatment B is administered at time steps 50, 70 and 90. c) Gene Configuration II and its response to the different treatments.

cancer drug was administered 3 times during the simulation at time steps 30, 50 and 70. In *Treatment B*, administration took place at time steps 50, 70 and 90. These experiments were carried out for the 2 gene configurations. Results suggest that, for *Gene Configuration I* (Fig. 1.b), the drug has a positive effect if administered earlier. Because in this system an over-proliferating oncogene has more chance of being activated first, as shown in previous work [1], eradicating cells initially has a direct impact. For *Gene configuration II* (Fig. 1.c), when cancer initiation is due to the loss of a tumour suppressor gene, this kind of treatment is more consistent regardless of the time of administration. Work is ongoing on a more detailed evaluation of the relationship between genome configuration and the success of different treatment strategies.

Models are needed to help explore the poorly understood role of chromosome missegregation in cancer progression and treatments [4]. Results show that cancer-like systems with different kinds of aneuploidy respond differently to cancer treatments. Work is ongoing to adapt the models to simulate other cancer treatments.

# References

[1] A. Araujo, P. Bentley, and B. Baum. Modelling the role of aneuplodiy in tumour evolution. In *Proceedings of Artificial Life XII (to appear)*, 2010.

[2] D. J. Baker, J. Chen, and J. M. A. van Deursen. The mitotic checkpoint in cancer and aging: what have mice taught us? *Current opinion in cell biology*, 17(6):583–589, 2005.

[3] R .T. Skeel. *Handbook of cancer chemotherapy*. Philadelphia: Lippincott Williams and Wilkins, 2003.

[4] Rocio Sotillo, Juan-Manuel Schvartzman, Nicholas D. Socci, and Robert Benezra. Mad2-induced chromosome instability leads to lung tumour relapse after oncogene withdrawal. *Nature*, 464:436–440, 2010.

# Agent-based Modelling of the Haematopoetic Cellular System

Daniel Jones, Mark d'Inverno, and Tim Blackwell

Goldsmiths, University of London, UK

The human vascular system is maintained through the self-organising behaviour of a vast population of cells, regulated through local interactions. Pathological states such as chronic myeloid leukaemia (CML) emerge from the accumulation of cellular malfunctions within this population, with dynamics which appear similar to that of selective evolution; however, biologists are still unclear exactly what conditions are required (or precluded) for this to occur.

Our work simulates and visualises these interactions, with the intention of generating testable hypotheses as to the causal networks which give rise to chronic cellular proliferation, as well as more typical behaviours such as cellular migration and homing. Adopting the agent-based paradigm, we argue, allows a biologist to accurately represent the *formation dynamics* of a process, encouraging a system-level conceptualisation of the domain and closing the cognitive gap between model and reality.

This poster portrays the trajectory of our research, outlining the motivations and methodologies for moving from an equation-based to agent-based paradigm with formal methods. We describe a new agent-based simulation framework, which harnesses GPU acceleration in order to radically increase the scale of system that we are able to model. Finally, we discuss our current work in validating simulation results based on clinical data.

CPSIA information can be obtained
at www.ICGtesting.com
Printed in the USA
BVHW041951161218
535687BV00013B/23/P